High Precision Infra-Red Stellar Interferometry

by

Benjamin F. Lane

ISBN: 1-58112-200-4

DISSERTATION.COM

USA • 2003

High Precision Infra-Red Stellar Interferometry

Dissertation.com
USA • 2003

ISBN: 1-58112-200-4
www.Dissertation.com/library/1122004a.htm

High-Precision Infra-Red Stellar Interferometry

Thesis by

Benjamin F. Lane

In Partial Fulfillment of the Requirements

for the Degree of

Doctor of Philosophy

California Institute of Technology

Pasadena, California

2003

(Defended Febuary 20th, 2003)

For my Father, who always wanted me to go to MIT.

Acknowledgements

First and foremost I wish to thank my advisor, Shri Kulkarni. Shri combines the tremendous experience, insight and energy of an outstanding scientist with the wisdom and patience of a great advisor, and I have benefited enormously from his guidance. For that I am truly grateful.

I also wish to thank Mark Colavita for all these years of friendship, advice, guidance and support. I can only hope to aspire to his level of technical brilliance and Zen patience.

I wish to thank the members of my committee, who have all at various times looked out for me: Bruce Murray for his wise counsel as well as lively discussions on many topics, Dave Stevenson for his theoretical insight and sage advice going as far back as my freshman days, and Mike Brown, Peter Goldreich and Geoff Blake who have all helped me understand the many aspects of the science I am doing.

The JPL crew deserves special mention; in particular, Andy Boden has helped me in more areas than I can count. Above all, he has taught me how to understand the statistics of measurements and experiments. I know this makes me a better scientist, and I am very grateful. Michelle Creech-Eakman has also helped me with much of my work, and been a friend always. Of course, PTI is the fruit of the labor of many people, all of whom deserve mention and gratitude: Bob, Chris, Dave, Gautam, Gerard, Kent, Mark S., Mike, Rachel, Robert, Sam, and many others. Among them, Dean Palmer is outstanding for keeping PTI alive despite my best efforts.

I wish to thank the staff of Palomar Observatory, in particular, Kevin Rykoski deserves special mention as it is his careful attention to the challenges of observing with PTI that has provided me with much of my data; he has truly gone above and beyond the call of duty. In addition, I have always looked forward to our nightlong wide-ranging discussions, when the sky is clear, the seeing sub-arcsecond and the interferometer humming along nicely; then life is good. I also wish to thank Jean for her observing skills, and Rose and Dipali for making such a good home away from home.

I know that through the years of graduate school I have received help from so many people at Caltech that it is difficult to list them all; but I am especially grateful to Shane, Antonin, Ian, Lori, Keith, Maciej, Maria, Eduardo, Matthew, Josh[2], Dave, Bryan, Edo and Ben. I would also like to thank the folks in planetary science for many valued discussions, and many enjoyable coffee hours. And of course, I am deeply grateful for the help from the planetary science office – especially the efforts of Donna, Irma and Ulrika.

I also wish to thank my close friends Marcus and Lauren. Graduate school is a long haul, and without their support and friendship I would surely have despaired of ever finishing.

I am grateful beyond measure to my mother, for her love and support, and for bringing me into this world. I have dedicated this work to my father, who never lived to see it. I can only hope it makes him proud.

Finally, this list cannot be complete without mention of my lovely, wonderful wife, Megan. Although this is hardly the place to express the full depth of my feelings for her, without any doubt she has made my life a joy and pleasure beyond words. Quite simply, I could not have done this without her.

Abstract

This dissertation describes work performed at the Palomar Testbed Interferometer (PTI) during 1998-2002. Using PTI, we developed a method to measure stellar angular diameters in the 1-3 milli-arcsecond range with a precision of better than 5%. Such diameter measurements were used to measure the mass-radius relations of several lower main sequence stars and hence verify model predictions for these stars. In addition, by measuring the changes in Cepheid angular diameters during the pulsational cycle and applying a Baade-Wesselink analysis we are able to derive the distances to two galactic Cepheids (η Aql & ζ Gem) with a precision of $\sim 10\%$; such distance determinations provide an independent calibration of the Cepheid period-luminosity relations that underpin current estimates of cosmic distance scales.

Second, we used PTI and the adaptive optics facility at the Keck Telescope on Mauna Kea to resolve the low mass binary systems BY Dra and GJ 569B, resulting in dynamical mass determinations for these systems. GJ 569B most likely contains at least one substellar component, and as such represents the first dynamical mass determination of a brown dwarf.

Finally, a new observing technique – dual star phase referencing – was developed and demonstrated at PTI. Phase referencing allows interferometric observations of stars previously too faint to observe, and is a prerequisite for large-scale interferometric astrometry programs such as the one planned for the Keck Interferometer; interferometric astrometry is a promising technique for the study of extra-solar planetary systems, particularly ones with long-period planets.

Contents

List of Tables

List of Figures

List of Acronyms

AO Adaptive Optics

BD Brown Dwarf

CCD Charge-Coupled Device

CHARA Center for High Angular Resolution Astronomy

COAST Cambridge Optical Aperture Synthesis Telescope

CT Constant Term (Metrology)

FT Fringe Tracker

GI2T Grand Interféromètre à 2 Télescopes

I2T Interféromètre á 2 Télescopes

IRAF Image Reduction & Analysis Facility

ISI Infrared Spatial Interferometer

JPL Jet Propulsion Laboratory

KI Keck Interferometer

LDL Long Delay Line

NIR Near Infra-Red

NPOI Navy Prototype Optical Interferometer

OPD Optical Path Delay

PTI Palomar Testbed Interferometer

RV Radial Velocity

SNR Signal-to-Noise Ratio

VLTI Very Large Telescope Interferometer

Chapter 1

Introduction to Interferometry

1.1 Why Interferometry?

In characterizing an astronomical telescope there are usually two principal figures of merit: light collecting power and angular resolution. The first is a function of the area of the primary mirror (scaling as $\sim D^2$, D being the telescope diameter), while the second is, according to basic diffraction theory, given by $\sim \lambda/D$, where λ is the wavelength of observation. As a consequence the best telescope is the one with the largest possible diameter; to date the largest operational telescope mirror has a diameter of 10 meters (m), with mirrors up to 30 m in diameter being in the planning stages. Although one might conceive of single telescopes with diameters in the range of 100 m or more, such a device is currently well beyond current technology. Thus if one desires very high angular resolution, it becomes necessary to resort to interferometry, i.e., the practice of combining light from two or more telescopes (separated by a baseline distance D) in such a way as to obtain an angular resolution of $\sim \lambda/D$, i.e., equivalent to that of a telescope of diameter equal to the separation between the telescopes. In this case D can be 100-1000 m, with correspondingly higher resolution.

At this point it should be noted that for a ground-based telescope the achievable angular resolution is usually not limited by diffraction but by the atmosphere; for such a telescope, turbulent variations in the atmospheric index of refraction distort the wavefront of the incoming starlight (making stars twinkle). In practice one can define a wavelength-dependent length scale (the Fried parameter, r_0) over which the incoming wavefront suffers less than 1 radian r.m.s. phase deviation. For a typical astronomical site $r_0 \sim 10$–20 cm for a wavelength of 0.5μm, increasing to ~ 50 cm at 2.2μm. The net result is to limit the angular resolution of a telescope, regardless of size, to λ/r_0, typically 0.5–1 arcsecond. However, there has in the past ten years been a great deal of progress in "adaptive optics" (AO), systems that use wavefront sensors and deformable mirrors to correct the effects of atmospheric turbulence in real time, and so effectively restore the diffraction-limited resolution of a large telescope. It is also worth mentioning two other techniques that can under limited circumstances provide diffraction limited observations: aperture masking and speckle interferometry. Nevertheless, these techniques can not do more than restore the diffraction limited performance of the telescope.

In principle, an interferometer is subject to the same atmospheric limitations as a single telescope. However, as we shall see, the principal effect of atmospheric seeing on an interferometer is to reduce its sensitivity rather than its resolution. Hence, while the atmosphere poses many practical challenges to interferometric observations, it does not fundamentally limit the high angular resolution we seek.

1.1.1 Resolving Single and Multiple Stars

How then is high angular resolution useful in astronomy, and what levels of resolution are required to be scientifically useful? One area where high spatial resolution techniques (e.g. interferometry and adaptive optics) are particularly valuable is in providing direct observational verification of astrophysical theory. Examples include measurement of such basic parameters as stellar radii and masses (by resolving binaries), circumstellar disk structures, and the structures of stellar atmospheres (e.g., via direct measurement of limb darkening). Such direct verification is required to bring current models from the typical $\sim 10\%$ uncertainties to the 1% level or better. Examples of this kind of work include measuring the mass-radius relation for low-mass stars (Lane, Boden & Kulkarni 2001, Chapter 2), measuring dynamical masses of low-metallicity evolved stars (Torres $et\ al.$ 2002) and low-mass stars (Boden & Lane 2000, Chapter 4), Cepheid distances and radii (Lane $et\ al.$ 2000, Chapter 3), measurements of the size and structure of disks around massive young stars (Eisner, Lane & Sargent 2003), and dynamical masses of brown dwarfs (Lane $et\ al.$ 2001, Chapter 6).

In considering the resolution required to resolve single stars it is instructive to consider the following argument: we assume that we are interested in blackbody emission from some astronomical source with temperature T, and we conduct our observations at wavelength

K-band Surface Brightness vs. Baseline

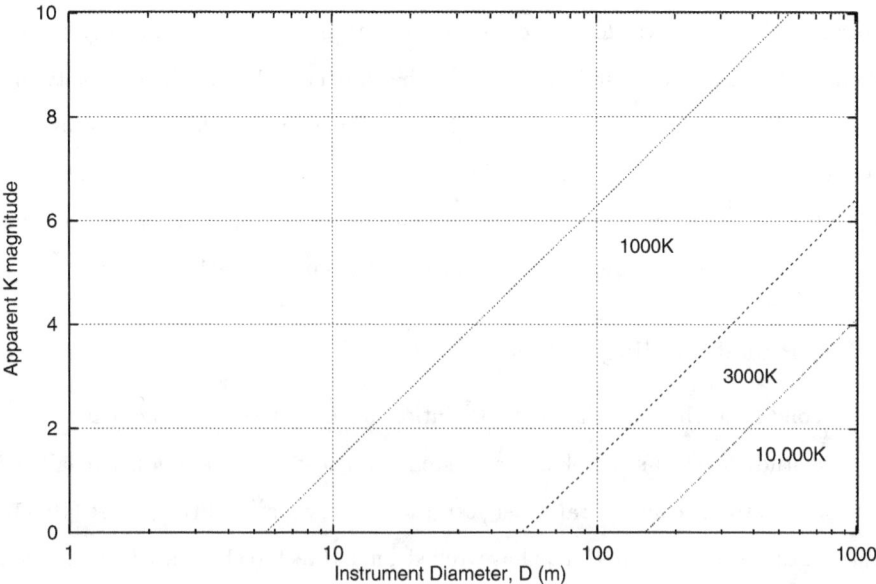

Figure 1.1: Corresponding surface temperature for a resolved source of a given K-band magnitude ($\lambda = 2.2\mu$m) and instrument diameter D.

λ with a telescope or interferometer with a characteristic size D. A source will have an apparent intensity I given by

$$I = B_\lambda(\lambda, T)\theta^2 \qquad (1.1)$$

where θ is the solid angle subtended by the source, and $B_\lambda(\lambda, T)$ is the Blackbody function. For the source to be resolved by our instrument we have

$$\theta \simeq \frac{\lambda}{D} \qquad (1.2)$$

Hence for a given apparent source intensity and telescope diameter we can determine the temperature of sources we can resolve; the result is shown in Fig. 1.1. Clearly, if we wish to resolve sources with temperatures typical of stars (10^3–10^4K), we will require baselines of 100 m or more. It is also noteworthy that resolved sources will be quite bright; however, this is fortunate, given the limited sensitivity of most interferometers (see below).

If we wish to resolve binary stars instead of single stars there are in principle visual binary stars with orbits that are resolvable with arcsecond-level imaging resolution. However, the orbital period ($P \propto a^{3/2}$) will for typical source distances of 1–100 pc be very large, up to hundreds or thousands of years. In addition, the associated line-of-sight velocity of the stars ($v_r \propto a^{-1/2}$) becomes very small ($<< 1$ kms^{-1}) making it difficult to measure the orbital velocity of such visual binaries via Doppler spectroscopy. Hence it is preferable to have the ability to resolve binaries with orbital semi-major axes of 1 AU or less; at typical source distances this implies a required angular resolution of $10^{-1} - 10^{-3}$ arcsec.

1.1.2 The Planet-Finding Problem

There is a second area where high angular resolution is particularly valuable, namely in the detection and study of extra-solar planetary systems. Such studies are uniquely challenging due to the often extreme contrast ratio between the primary star and the planet (10^3–10^9). In practice most workers in this area have relied on indirect techniques (radial velocity measurement of the reflex motion of the parent star; Mayor & Queloz, 1995), or made use of a known modulation of the planetary signal, i.e., transits (HD 209458, Charbonneau et al. 2000). However, the high angular resolution available with interferometry may allow other detection methods, the details of which depend on the type of planet one is searching for. In addition to detection, the use of interferometry is particularly interesting in that it promises more detailed information than otherwise obtainable, i.e., direct mass measurements and eventually spectroscopy. Below we outline three types of planets that can be studied with interferometry, the techniques themselves will be discussed in section 1.5.

1.1.2.1 Hot Jupiters

The recent discovery of massive planets in short-period orbits has sparked a thriving industry in planet detection. The first such planet discovered (Mayor & Queloz, 1995), 51 Peg, has a minimum mass of 0.45 M_J and orbits a mere 0.051 AU from its parent star (Fig. 1.3). The presence of such a massive planet in such a close orbit had not been expected, and indeed posed a challenge to formation theorists.

Since the initial discovery, some 100 more planets have been found, principally by radial velocity techniques. A large fraction of the systems found to date ($\sim 25\%$) orbit within 0.07 AU of their parent star, although this is very likely a selection effect due to the observing

1 mas

Figure 1.2: The typical Hot Jupiter system (to scale!). 1mas $= 10^{-3}$ arcseconds

technique. For the purposes of this discussion I will call these massive, close planets "Hot Jupiters." At present the preferred formation scenario for the Hot Jupiters is that they form at a large distance ($\sim$ 5AU, beyond the "snow line") in a manner similar to the formation of Jupiter. A transport mechanism is then invoked to explain how they migrate in to closer orbits, e.g., tidal coupling with the proto-planetary disk.

For the purposes of this discussion let us consider a system consisting of a G0V star at a distance of 15 pc, with a $\sim 1M_J$ planet in a 0.05 AU orbit. We take the radius of the planet to be $\sim 1R_J$, consistent with observed transits, and expected by theory. The effective temperature of the planet will be determined principally by radiative equilibrium with the parent star:

$$T = \left[\frac{(1-A)R_*^2}{4a^2} \right]^{\frac{1}{4}} T_* \qquad (1.3)$$

For our prototypical system this gives $T \sim 1000$K.

The result is shown in Fig. 1.3. The high contrast ratio is the main challenge to directly detecting these types of planets. Note that the angular separation of the star and planet is 5 mas (0.05 AU / 15 pc = 0.0033 arcsec). Thus if we wish to separate the light of the planet from that of the star we will require an angular resolution of this order. For operation at 2 μm, this implies an aperture diameter of $D = \lambda/\theta \sim 100$ m. This is the motivation for using interferometry for the study of these systems, specifically by using differential interferometry (Sec.1.5.3).

1.1.2.2 Outer System Planets

In 1999 it was announced that the system υ Andromeda contained at least 3 planets (Butler et al. 1999) , including one in a 0.059 AU orbit. One of the most interesting aspects of this

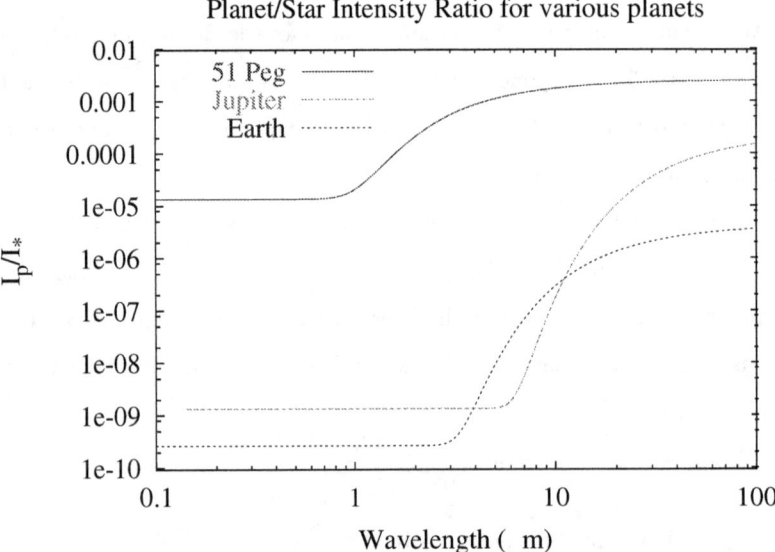

Figure 1.3: Calculated intensity ratio for three types of planets: Earth, Jupiter and a "Hot Jupiter," i.e., a 1 R_J planet at 0.05 AU from a G2V star. The plot includes contributions from emitted (the planet being in radiative equilibrium for its distance) and scattered light. The assumed albedo was 0.5.

system concerns the long-term stability of this system. Models have shown that unless the relative inclinations of the objects is greater than ~ 20 degrees, the system in dynamically unstable on timescales less than a few Gyr (Laughlin & Adams 1999). The uncertainty in the models comes from that fact that the inclinations (and thus also the actual masses) of the planets is unknown. This is because the radial velocity techniques only measure $M \sin i$ and cannot determine inclination. Clearly a method of measuring the inclination of these systems is desirable.

For massive planets in outer system orbits (a > 1 AU) the star-planet intensity ratios can be substantial, making direct detection difficult. However, in this case astrometric techniques become increasingly useful. The astrometric reflex motion of the parent star will be given by

$$\Delta\theta \simeq 1000 \frac{M_p}{M_*} \frac{a_p}{d} \mu\text{arcsec} \tag{1.4}$$

where M_p is the mass of the planet in Jupiter masses, M_* is the mass of the central star in solar masses, a_p is the orbital semi-major axis in AU and d is the distance to the system in parsecs. As the best available (non-interferometric) astrometry to date is on the order of 1 mas we see that detecting planets using astrometry is very challenging. However, the scientific payoff is high as it is more sensitive than radial velocity techniques at large orbital distances, and in addition it directly provides the orbital inclination, eliminating the mass uncertainty inherent in RV measurements. As we shall see, interferometric astrometry is expected to achieve precisions of 20–50 μarcseconds from the ground, and 1–5 μarcseconds with space-based instruments.

1.1.2.3   Earths

Terrestrial planets are particularly difficult to study, as they lack both the mass to produce an appreciable astrometric ($\leq 1\mu$ arcsec) or Doppler signature (~ 0.1 ms^{-1}), are small enough to have a very high contrast ratio (10^9), and are far enough from their parent star that the probability of a transit is quite small (10^{-3}) and shallow (μmag). Nevertheless, there are a number of techniques that are being developed to detect such planets, including interferometric and coronographic, as well as transit approaches. These systems all require space-based observations, and so fall well outside the scope of this work. Nevertheless, advanced interferometric techniques such as nulling, if they can be demonstrated on the

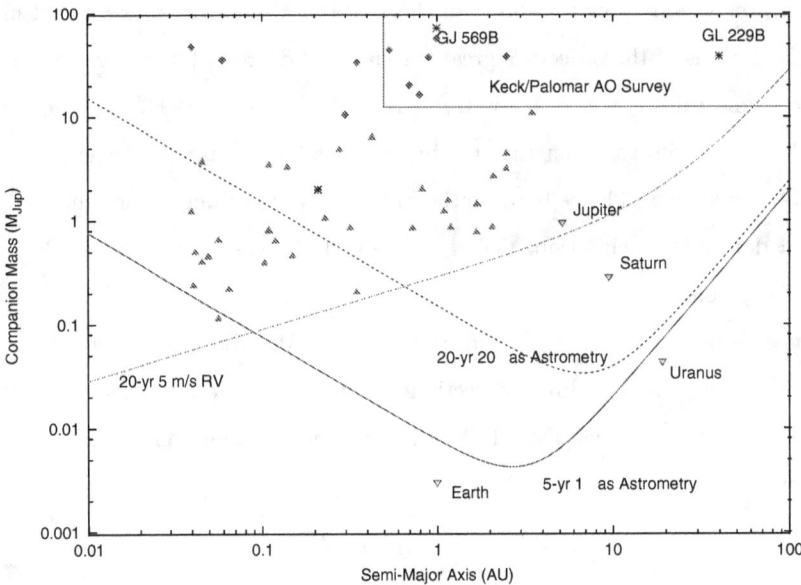

Figure 1.4: Calculated astrometric signatures of a number of Solar system and extra-solar planets. Also indicated is the phase space sampled by various indirect techniques, i.e., radial velocity monitoring and astrometry.

ground, will likely be used for such studies in the next decade.

1.1.3 Probing Circumstellar Environments

The high angular resolution available with stellar interferometry allows one to study the close circumstellar environment around a range of stars. Perhaps most interestingly is the study of star and planetary formation, where observations at PTI (Eisner et al. 2002) and elsewhere have resolved circumstellar structure on the sub-AU scale around a number of Herbig AeBe and T Tau systems.

1.2 Theory of Interferometry

I will start with a very simple interferometer in order to illustrate the basic concepts of interference fringes, fringe visibility and phase, and angular resolution. Then I review the relationship between a source of interest and the observables, including relating fringe visibility to single and double stars. For a more thorough introduction to the theory of

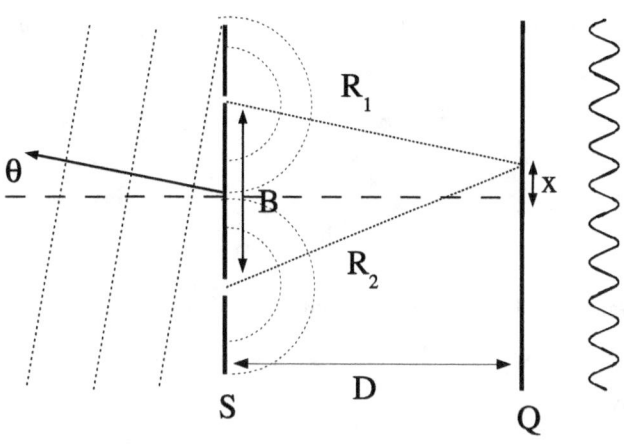

Figure 1.5: A double-slit experiment.

interferometry, the reader is referred to Mandel & Wolf (1995) or Perley et al. (1989).

1.2.1 The Double-slit Experiment

To illustrate the basic interference phenomenon that underlies interferometry, consider the experimental setup in Fig. 1.5: monochromatic light comes from the left at an angle θ to the horizontal axis and encounters a screen S, with two pinhole apertures separated by a distance B. The Hyugens-Fresnel principle [a] allows us to treat the slits as sources of spherical waves. At any point x on the image plane Q, the amplitude of the resulting wave is given by adding up the contributions from the slits (ignoring time-dependence)

$$E_1 \sim e^{i\frac{2\pi}{\lambda}R_1(x)} \tag{1.5}$$

$$E_2 \sim e^{i\frac{2\pi}{\lambda}(R_2(x)+B\sin(\theta))} \tag{1.6}$$

at location x (see Fig. 1.5) the field amplitude will be $\sim E_1(x) + E_2(x)$ and the measured intensity, where $I \equiv \langle EE^* \rangle/2$, and assuming equal-sized apertures, is

$$I = (1 + \cos(\frac{2\pi}{\lambda}\delta)) \tag{1.7}$$

---

[a]At a given instant in time, every unobstructed point in a wavefront serves as a source of secondary spherical waves of the same frequency as the original wave. The amplitude of the field at any subsequent point is the superposition of all such secondary waves.

where δ is the total (before and after the screen) difference in path-length of the two beams, i.e.,

$$\delta = (R_1 - R_2) - B\sin(\theta) \tag{1.8}$$

For $B \ll D$

$$R_1 - R_2 \sim x\frac{B}{D} \tag{1.9}$$

and thus the intensity in the image plane will vary sinusoidally between 0 and 2, with a phase that is related to the angle of the incoming wavefront, θ. Such a pattern is known as an *interference fringe*.

Now, consider the effect of adding a second source, producing a wavefront with an angle θ'; this will result in a second interference pattern with a slightly different phase. If the sources are incoherent, the resulting superposition at Q is

$$
\begin{aligned}
I_2 &= I' + I \tag{1.10}\\
&= 2\left(1 + \cos(\frac{\pi}{\lambda}(\delta' - \delta))\cos(\frac{\pi}{\lambda}(\delta' + \delta))\right) \tag{1.11}
\end{aligned}
$$

thus the fringe pattern remains, but is less sharp (i.e., no longer goes to 0). At this point it is useful to introduce the concept of the fringe contrast or *visibility*, defined by A. Michelson as

$$V = \frac{I_{max} - I_{min}}{I_{max} + I_{min}} \tag{1.12}$$

Clearly $0 \le V \le 1$. For the example above,

$$V = \cos(\frac{\pi}{\lambda}(\delta' - \delta)) \tag{1.13}$$

and hence one can infer the presence of a second emission source when the measured fringe visibility differs substantially from unity. In common practice, the *angular resolution* is the angle $\Delta\theta = \theta' - \theta$ for which the fringes disappear ($V = 0 \Rightarrow \delta' - \delta = \lambda/2$). For $\theta \sim 0$ we have

$$\Delta\theta = \frac{\lambda}{2B} \tag{1.14}$$

Note that the exact expression for the angular resolution can differ by factors of order unity for various instruments and definitions of resolution.

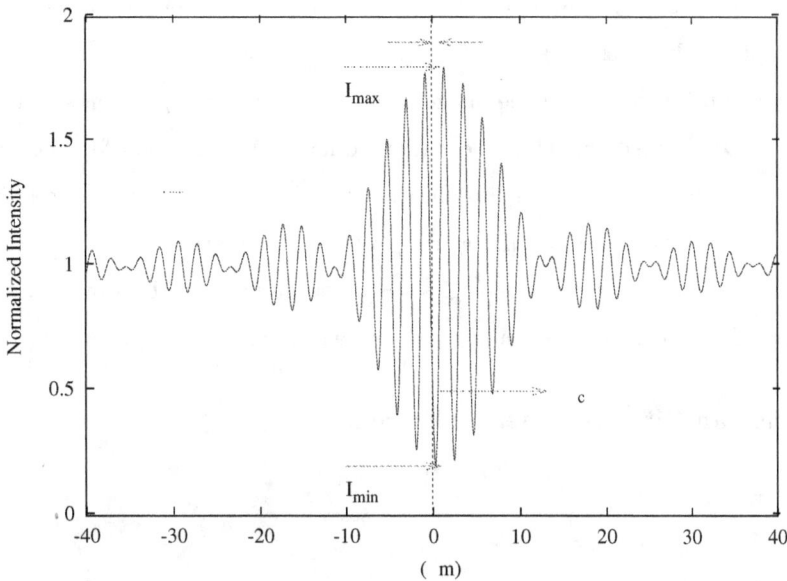

Figure 1.6: A model fringe, based on Eqn. 1.18, showing the definitions of the various parameters. In this case $\lambda = 2.2\mu$ m, $\Delta\lambda = 0.4\mu$ m, and $V = 0.8$. The fringe phase ϕ is usually defined as the shift from zero relative path difference.

If instead of monochromatic light we illuminate the double-slit experiment with broadband light, the resulting fringe pattern can be calculated by linear addition of the frequency components. Assuming an illumination with frequency centered at ν_0 and bandwidth $\Delta\nu$, with constant intensity per unit frequency I_ν, the resulting fringe pattern is

$$I_{Broadband} = \int_{\nu_0-\Delta\nu/2}^{\nu_0+\Delta\nu/2} I_\nu \left(1 + \cos\left(2\pi\frac{\nu}{c}\delta\right)\right) d\nu \tag{1.15}$$

$$= I_\nu \left[\nu + \frac{\sin(2\pi\delta\nu)}{2\pi\delta/c}\right]_{\nu_0-\Delta\nu/2}^{\nu_0+\Delta\nu/2} \tag{1.16}$$

$$= I_\nu\Delta\nu \left(1 + \frac{\sin(\pi\delta\Delta\nu/c)}{\pi\delta\Delta\nu/c}\cos(2\pi\delta\nu_0/c)\right) \tag{1.17}$$

or equivalently

$$I_{Broadband} = I_\lambda\Delta\lambda \left(1 + \frac{\sin(\pi\delta/\Lambda_c)}{\pi\delta/\Lambda_c}\cos(2\pi\delta/\lambda_0)\right) \tag{1.18}$$

where $\Lambda_c \equiv \lambda_0^2/\Delta\lambda$ is the *coherence length*. The fringe pattern remains, but is now modulated by an envelope function given by the cosine-transform of the bandpass function. For large δ, equivalent to being far from the instrument centerline, the fringes disappear.

If one considers an interferometer as a glorified double-slit experiment, where the pin-holes are replaced by telescopes, the free-space propagation between S and Q may be replaced by an optical relay, and the image plane is sampled by a position-sensitive detector (e.g. a CCD array), it should be clear that we now have a method for calculating the response of a double-slit experiment to one or more distant broadband sources, based on the phase and contrast of the resulting fringe pattern. However, it may not be obvious how to relate complex source morphologies to the fringe pattern. Fortunately, there is a straightforward relationship between the two, as we explore in the next section.

1.2.2 The van Cittert-Zerneke Theorem

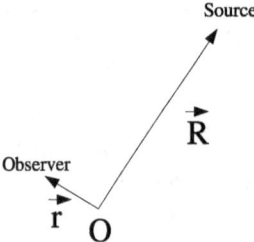

Figure 1.7: An observer located at $\vec{r}$ observes a source at $\vec{R}$.

Consider the situation typically encountered in astronomy, where an observer seeks to learn about a distant source (Fig. 1.7). Because it is so far away, it is reasonable to treat the source as if it were two-dimensional, i.e., projected onto a distant "celestial sphere" of radius $|\vec{R}|$; in this case the source can be seen if it possesses a time-variable electric field $\mathcal{E}_\nu(\vec{R})$, which results in the emission of electromagnetic radiation that propagates to $\vec{r}$ where it can be detected[b]. As in the previous section, Huygens Principle provides an easy way to calculate the resulting field

$$E_\nu(\vec{r}) = \int \mathcal{E}_\nu(\vec{R}) \frac{e^{2\pi i \nu |\vec{R}-\vec{r}|/c}}{|\vec{R}-\vec{r}|} dS \qquad (1.19)$$

in the case of interferometry we are interested in measuring the spatial correlation of $E_\nu(\vec{r})$;

---

[b]We ignore polarization effects for simplicity, and also note that the linear nature of Maxwells equations in this regime allows one to simply add up contributions from all frequencies linearly.

to do so we make use of the mutual coherence function

$$V_\nu(\vec{r}_1, \vec{r}_2) = \langle E_\nu(\vec{r}_1) E_\nu^*(\vec{r}_2) \rangle \tag{1.20}$$

and hence

$$V_\nu(\vec{r}_1, \vec{r}_2) = \left\langle \int \int \mathcal{E}_\nu(\vec{R}_1) \mathcal{E}_\nu(\vec{R}_2) \frac{e^{2\pi i \nu |\vec{R}_1 - \vec{r}_1|/c}}{|\vec{R}_1 - \vec{r}_1|} \frac{e^{-2\pi i \nu |\vec{R}_2 - \vec{r}_2|/c}}{|\vec{R}_2 - \vec{r}_2|} dS_1 dS_2 \right\rangle \tag{1.21}$$

If we assume that the source emission is spatially incoherent, i.e.,

$$\langle \mathcal{E}_\nu(\vec{r}_1) \mathcal{E}_\nu^*(\vec{r}_2) \rangle = \langle |\mathcal{E}_\nu(\vec{r}_1)|^2 \rangle \delta(\vec{r}_1 - \vec{r}_2) \tag{1.22}$$

and furthermore define the source intensity $I_\nu(\vec{s}) = \langle |\mathcal{E}_\nu(\vec{s})|^2 \rangle$, where $\vec{s} = \vec{R}/|\vec{R}|$, by interchanging integrals and neglecting second-order terms we find

$$V_\nu(\vec{r}_1, \vec{r}_2) \sim \int I_\nu(\vec{s}) e^{-2\pi i \nu \vec{s} \cdot (\vec{r}_1 - \vec{r}_2)} d\Omega \tag{1.23}$$

Eqn. 1.23 is known as the van Cittert-Zerneke theorem, and it relates the source intensity distribution and the the fringe visibility measured by an interferometer via a Fourier transform. In typical practice the above quantity is normalized such that

$$\hat{V}_\lambda(\vec{B}) = \frac{\int I_\lambda(\vec{s}) e^{-\frac{2\pi i}{\lambda} \vec{s} \cdot \vec{B}} d\Omega}{\int I(\vec{s}) d\Omega} \tag{1.24}$$

where λ is the wavelength of observation and $\vec{B} = \vec{r}_1 - \vec{r}_2$ is referred to as the interferometric baseline. Note that $\hat{V}_\lambda(\vec{B})$ is a complex quantity; its amplitude is equivalent to the fringe visibility defined by Eqn. 1.12, while the phase of the measured fringe pattern depends on instrumental path-lengths (c.f. δ) as well as the source morphology and instrumental geometry. Commonly, the baseline components are referred to in terms of spatial frequencies and the uv-plane

$$u \equiv \frac{B_x}{\lambda} \tag{1.25}$$

$$v \equiv \frac{B_y}{\lambda} \tag{1.26}$$

1.2.3   Fringe Visibility, Imaging and Models

Given Eqn. 1.24 one might imagine measuring $\hat{V}_\lambda(\vec{B})$ for all possible baseline orientations and lengths (at least up to some maximum baseline), and then recovering the source image (up to a limiting resolution $\sim \lambda/B$) via an inverse Fourier transform. This is in effect what a telescope does. However, with an interferometer it is usually impossible to fully sample the uv-plane, and thus the recovered image will contain artifacts from the sampling function. Specifically, assume a sampling function for i measurements in the uv-plane

$$S(\vec{s}) = \sum_i \delta(\vec{s_i}) \tag{1.27}$$

the measured visibility function is

$$V_\lambda(\vec{B}) = \hat{V}_\lambda(\vec{B})S(\vec{s}) \tag{1.28}$$

and by the convolution theorem, the inverse Fourier transform produces

$$\tilde{I}_\lambda(\vec{s}) \;\;=\;\; IFT[\hat{V}_\lambda(\vec{B})S(\vec{s})] \tag{1.29}$$

$$=\;\; I_\lambda(\vec{s}) * IFT[S(\vec{s})] \tag{1.30}$$

Numerous techniques have been developed to deconvolve the resulting "dirty image." However, in the case of very limited uv-plane coverage, such as is the case with current optical interferometers, it is usually preferable to forward-model the visibility function based on an assumed parametric source model, then adjust the model parameters (e.g. via least-squares techniques) to fit the observed visibilities. Below we derive the visibility functions for the most common interferometric targets, single and binary stars.

1.2.3.1   Single Stars

For a single point source $I(\vec{s}) = I_0\delta(\vec{s})$ Eqn. 1.24 implies unit visibility. However, most stars are not perfect point sources, but have some finite angular diameter, θ. If we assume circular symmetry, the Fourier transform can be converted into a Hankel transform, i.e.,

$$\hat{V}_\lambda(\vec{B}) \;\;=\;\; \frac{\int I_\lambda(\vec{s})e^{-\frac{2\pi i}{\lambda}\vec{s}\cdot\vec{B}}d\Omega}{\int I(\vec{s})d\Omega} \tag{1.31}$$

$$= \frac{\int_0^{2\pi} d\phi \int_0^{\infty} dr I_\lambda(r) r e^{-\frac{2\pi i}{\lambda} r B \cos\phi}}{\int_0^{2\pi} d\phi \int_0^{\infty} dr I_\lambda(r) r} \qquad (1.32)$$

$$= \frac{\int_0^{\infty} dr I_\lambda(r) r J_0(xr)}{\int_0^{\infty} dr I_\lambda(r) r} \qquad (1.33)$$

where J_n is the nth order Bessel function and

$$x = \frac{\pi B \theta}{\lambda} \qquad (1.34)$$

The Hankel transforms have been tabulated, and thus we can use the following relation

$$\int_0^1 (1 - r^2)^\nu r J_0(rx) dr = 2^\nu \Gamma(\nu + 1) \frac{J_{\nu+1}(x)}{x^{\nu+1}} \qquad (1.35)$$

to derive the visibility function for a uniform disk model ($\nu = 0$)

$$\hat{V}_\lambda(\vec{B}) = \frac{\int_0^1 I_\lambda r J_0(rx) dr}{\int_0^1 I_\lambda r dr} \qquad (1.36)$$

$$= 2\frac{J_1(x)}{x} \qquad (1.37)$$

$$= 2\frac{J_1(\pi B \theta / \lambda)}{\pi B \theta / \lambda} \qquad (1.38)$$

If instead of a uniform disk model we wish to consider the effect of stellar limb darkening, it is often the case that stellar intensity profiles are given as a function of $\mu = \cos(\theta)$ where θ is the azimuth of a surface element of the star. Thus $\mu = \sqrt{1 - r^2}$ and for the case of a "fully limb darkened" model, $I_\lambda(\mu) = \mu \Rightarrow \nu = 1/2$, giving

$$\hat{V}_\lambda(\vec{B}) = \frac{\int_0^1 \sqrt{1 - r^2} r J_0(rx) dr}{\int_0^1 \sqrt{1 - r^2} r dr} \qquad (1.39)$$

$$= 3\sqrt{\frac{\pi}{2}} \frac{J_{3/2}(x)}{x^{3/2}} \qquad (1.40)$$

Usually one encounters limb darkening models that lie between the uniform disk and fully darkened disk models, parameterized by a limb darkening coefficient u_λ such that

$$I_\lambda(\mu) = 1 - u_\lambda(1 - \mu) \qquad (1.41)$$

the resulting visibility function is

$$\hat{V}_\lambda(\vec{B}) = \frac{(1 - u_\lambda) \int_0^1 r J_0(xr)dr + u_\lambda \int_0^1 \sqrt{1 - r^2} r J_0(xr)dr}{(1 - u_\lambda) \int_0^1 r dr + u_\lambda \int_0^1 \sqrt{1 - r^2} r dr} \qquad (1.42)$$

$$= \frac{(1 - u_\lambda)\frac{J_1(x)}{x} + u_\lambda \sqrt{\frac{\pi}{2}} \frac{J_{3/2}(x)}{x^{3/2}}}{(1 - u_\lambda)/2 + u_\lambda/3} \qquad (1.43)$$

For interferometric operations near or slightly below the nominal angular resolution there is a degeneracy between the angular size and the limb darkening (Fig. 1.8), and hence it is important to have a good model atmosphere available if one wishes to determine stellar angular diameters to high precision. Conversely, given a suitable long interferometric baseline it becomes possible to measure the limb-darkening directly, and hence provide an empirical measure of the atmospheric properties of a range of stars.

Given the above degeneracy it is usually the case that for observations at a spatial frequency below the resolution limit one can fit a uniform disk model to the data, derive a uniform disk diameter θ_{UD}, then convert it to a limb darkened diameter θ_{LD} using a simple correction formula that depends only on the limb darkening parameters (see Chapter 3).

1.2.3.2 Binary Stars

In the case of a binary star, as long as the individual stellar components are unresolved by the interferometer, I can be modeled as two point sources located at $\vec{s}_0$ and $\vec{s}_1$.

$$I(\vec{s}) = I_0\delta(\vec{s} - \vec{s}_0) + I_1\delta(\vec{s} - \vec{s}_1) \qquad (1.44)$$

Defining the intensity ratio $R = I_1/I_0$ and $\Delta\vec{s} = \vec{s}_1 - \vec{s}_0$, we find

$$\hat{V} = \frac{1}{1 + R}\left(1 + R\, e^{-\frac{2\pi i}{\lambda}\Delta\vec{s}\cdot\vec{B}}\right) \qquad (1.45)$$

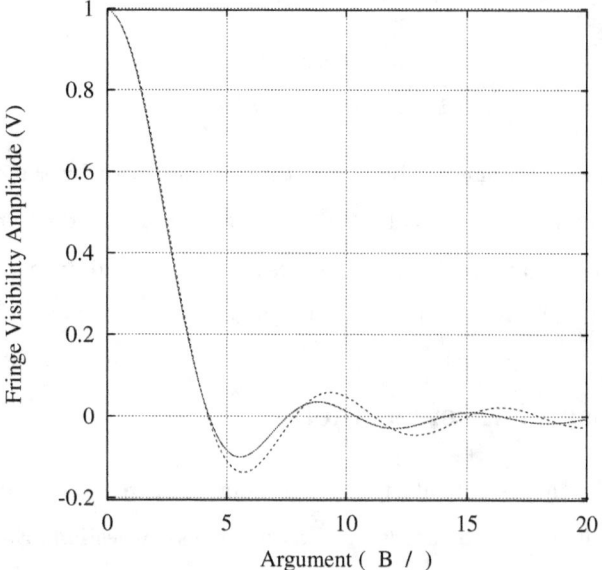

Figure 1.8: Plotted is the fringe amplitude for a uniform disk (dashed) and a 10% larger limb-darkened (solid) disk model of a single star. Only at long baselines is the degeneracy between size and limb darkening broken.

It is useful to write this visibility in a slightly different form

$$\bar{V} = |\hat{V}|e^{i\phi} \tag{1.46}$$

where

$$|\hat{V}|^2 = \frac{1}{(1+R)^2}\left(1 + R^2 + 2R\cos(\frac{2\pi}{\lambda}\Delta\vec{s}\cdot\vec{B})\right) \tag{1.47}$$

The quantity $|V|^2$ (or simply V^2) is usually what is measured by an optical interferometer; it corresponds to the contrast of the observed fringes.

In the case of a binary star system with partially resolved stellar disks, it should be clear from the linearity of the van Cittert-Zerneke theorem that the visibilities are combined such that

$$|\hat{V}|^2 = \frac{1}{(1+R)^2}\left(V_1^2 + V_2^2 R^2 + 2V_1 V_2 R\cos(\frac{2\pi}{\lambda}\Delta\vec{s}\cdot\vec{B})\right) \tag{1.48}$$

where V_i is the uniform-disk visibility of the ith component from Eqn. 1.36.

The phase of the complex visibility of a binary star is given by

$$\phi = \arctan\left(\frac{R\sin(\frac{2\pi}{\lambda}\Delta\vec{s}\cdot\vec{B})}{1 + R\cos(\frac{2\pi}{\lambda}\Delta\vec{s}\cdot\vec{B})}\right) \tag{1.49}$$

However, the phase measured by a single pair of apertures is corrupted by the effects of atmospheric turbulence on very short timescales (in the optical and infrared ~ 1 radian in ~ 10 ms) and contains no useful information, unless it is measured simultaneously with a phase reference source; see the discussion of advanced techniques.

1.3 A Brief History of Interferometry

The first person[c] to suggest the use of interference phenomena to measure the angular diameters of stars was H. Fizeau (Fizeau 1868). Based on his suggestion, E. Stéphan obtained the first interferometric measurements of stars in 1872 (Stéphan 1873, 1874). Using masked 50 and 65 cm apertures, he concluded that the stars must all have very small angular sizes based on the fact that he consistently observed sharp interference fringes.

The theory of interference was explored mathematically by A. Michelson around 1890, who also used interferometry to measure the Galilean moons of Jupiter around this time (1891). The first binary star was resolved with interferometry in 1896 (Schwarzschild 1896), followed by A. Michelsons observation of Betelgeuse, for which an angular diameter was obtained (Michelson & Pease 1921). Michelson used the 100-inch telescope on Mt. Wilson, later outfitted with a 20-foot baseline extender for his observations on a number of single and binary stars.

With the success of the Mt. Wilson 20-foot interferometer, F. G. Pease, who had collaborated with Michelson, attempted to build a larger standalone instrument with a 50-foot baseline. Unfortunately he was pushing the limits of available technology, and the 50-foot interferometer was never successful. Optical stellar interferometry seemed relegated to the dustbin of history.

However, with the rapid advancement of radio interferometry after WW II, a novel idea, intensity interferometry, was developed by Hanbury-Brown and Twiss (Hanbury-Brown & Twiss 1956). It was successfully applied at the Narribri interferometer in the 1960's and used

---

[c]For a complete and interesting history of interferometry, see Lawson, 2001. A very useful compilation of fundamental papers in this area is also available in Lawson, 1997.

to measure the angular diameters of a number of early-type stars. Unfortunately, intensity interferometry avoids the exacting requirements of path-length control at the expense of a severe sensitivity penalty, and is limited to only the brightest stars. Nevertheless, the Narribri observations provided an extremely valuable observational foundation for stellar models that are used to this day.

Beginning in the 1970's, A. Labeyrie (1975) first combined light from separated telescopes and built the first modern stellar interferometer. The crucial technologies were fast electronic detectors, digital control computers, piezoelectric actuators and laser metrology systems. In the late 1970's a program at MIT was initiated (Shao & Staelin 1980) with the goal of developing a fringe-tracking interferometer for astrometry. This effort led directly to the Mk I–III family of interferometers, as well as providing design heritage for the NPOI, PTI and ultimately the Keck Interferometer.

The late 1980's and early 1990's saw a great deal of development in interferometry, with a number of interferometers coming on-line. These tended to be single-r_0 instruments with 2–4 apertures; however, they provided experience and motivation for further development, as well as a steadily growing stream of scientific results (Fig. 1.9). One notable change has been a general migration toward operation in the near-IR, where atmospheric constraints are significantly relaxed compared to operation in the visible regime; this has been made possible by the advent of high-performance detectors in this wavelength region. Toward the end of the decade, 6-aperture combination was achieved at NPOI.

Interferometry has, as of early 2003, bright prospects: two large ground-based arrays are under construction (the Keck Interferometer and the VLTI[d]), as well as several other systems (e.g., CHARA). In addition NASA and ESA have several space-based interferometers in various stages of planning and development, most notably the Space Interferometer Mission, intended for microarcsecond-level (μas) astrometry, and the future Terrestrial Planet Finder.

1.4 Practice of Interferometry

In order to achieve the high angular resolution, one would expect based on the discussion in Sec. 1.2 there are a number of practical hurdles to overcome. First one has to collect portions

[d]Very Large Telescope Interferometer

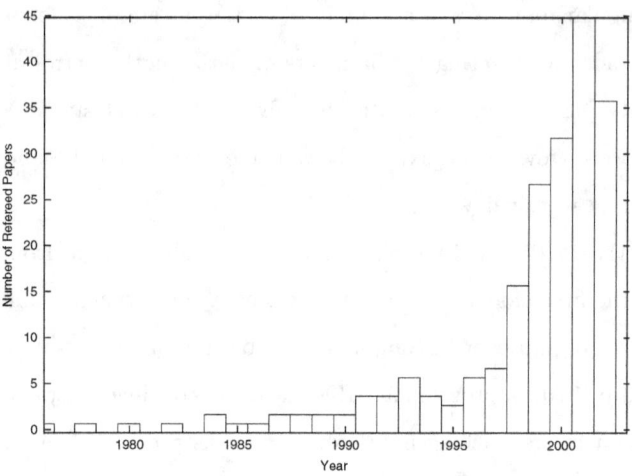

Figure 1.9: Recent publications based on interferometric results, as a function of time. (Source: ADS abstract service.)

of the incoming wavefront in two or more apertures, typically telescopes or siderostats are used. Second, in a direct-detection interferometer, the light has to be transported to a central location where it is combined. Third, the path-lengths have to be equalized, and the effects of atmospheric turbulence compensated or followed. This is usually done by adding variable amount of optical path delay (OPD) into the arms of the interferometer, although there exist designs where the apertures or beam combiner is moved so as to provide the path-length compensation (I2T, GI2T). Fourth, the beams must be combined and the fringe parameters measured. Finally, all of these subsystems must be controlled and coordinated in real time, usually by a complex assortment of computers and servo systems.

However, for ground-based instrument probably the most important consideration is how to correct for the effects of atmospheric turbulence. Below I discuss the phenomenon of atmospheric turbulence and how it impacts and interferometer, then I will briefly outline the practical aspects of designing an interferometer.

1.4.1 Atmospheric Turbulence and Seeing

The atmosphere above an interferometer can be characterized by the Reynolds number, $Re = VL/\nu$, where V is a characteristic velocity, L a characteristic length scale, and ν is the kinematic viscosity of air ($\sim 1.5 \times 10^5 \, \mathrm{m^2 s^{-1}}$). For length scales above a few meters and

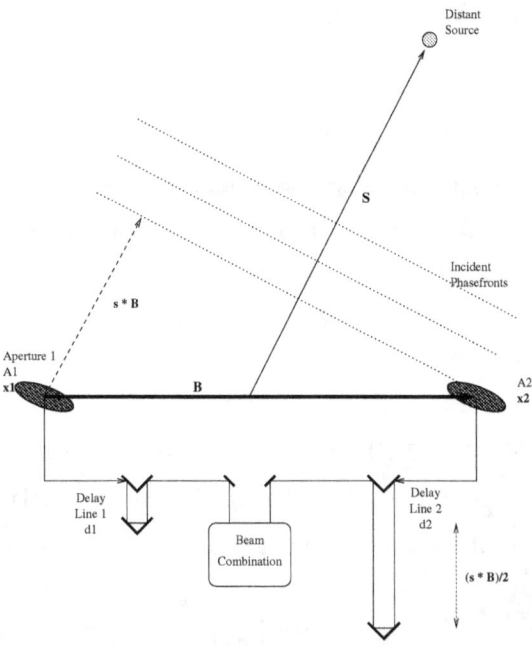

Figure 1.10: A schematic of a simple one-baseline interferometer. (From Lawson 2000.)

typical wind speeds of a few ms^{-1} we have Re$\sim 10^6$, which implies that the flow is turbulent (Faber 1995). The turbulent eddies correspond to parcels of air at different temperatures, which because they are in pressure equilibrium necessarily have different densities, and hence indices of refraction. The net effect is that the starlight must propagate through a medium with a rapidly varying (spatially and temporally) index of refraction; this causes distortions in the wavefront and makes the resulting image both blurry and variable.

The turbulence can be characterized by a distance over which the wavefront remains flat. This distance, known as the Fried parameter or the atmospheric coherence length, is given by

$$r_0 = \left[0.423 k^2 \sec \phi \int C_n^2(z) dz \right]^{-3/5} \tag{1.50}$$

where $k = 2\pi/\lambda$ is the wavenumber of the light, ϕ is the zenith angle of the observation, and C_n^2 is a parameter representing the turbulence amplitude (the structure constant). For typical observing sites, we have

$$r_0 \simeq 0.1 \left(\frac{\lambda}{0.5\mu m}\right)^{6/5} m \qquad (1.51)$$

At this point one can relate the coherence length to time by treating the atmospheric distortions as a fixed screen, blowing past the aperture with some wind speed, v; this is the frozen-flow hypothesis, and it implies

$$\tau_0 = \frac{r_0}{v} \qquad (1.52)$$

for typical wind speeds $\sim 10ms^{-1}$ this gives $\tau_0 \sim 10$ ms at 0.5 μm.

The phase fluctuations introduced by the atmosphere are correlated over small angles, characterized by the angular isoplanatic angle θ_0 such that

$$\left\langle \sigma_\theta^2 \right\rangle = \left(\frac{\theta}{\theta_0}\right)^{5/3} \qquad (1.53)$$

where

$$\theta_0 = \left[2.914k^2 (\sec \phi)^{8/3} \int C_n^2(z) z^{5/3} dz\right]^{-3/5} \qquad (1.54)$$

From the definition of the Fried parameter (Eqn. 1.50) we find that

$$\theta_0 = 0.314 \cos(\phi)\frac{r_0}{H} \qquad (1.55)$$

where H is the mean effective turbulence height

$$H \equiv \left[\frac{\int C_n^2(z) z^{5/3} dz}{\int C_n^2(z) dz}\right]^{3/5} \qquad (1.56)$$

In order to sample an undistorted portion of the wavefront it is necessary to collect ~ 100 photons per "coherence volume" ($\tau_0 r_0^2$). At the PTI the atmospheric coherence time (τ_0) is 10-20 ms, and the coherence length in the K-band (2.2 μm) is on the order of 40 cm. With an instrumental efficiency of only a few percent (due mostly to the many mirrors in the beam-path), the result is an effective tracking limit of 7th magnitude – clearly a significant limitation to the instrument.

1.4.2 A Basic Interferometer

Most of the currently operational interferometers have a common set of subsystems, outlined below. As most of the work discussed in this thesis was done at the Palomar Testbed Interferometer, as I outline the typical interferometer I will also provide a specific description of PTI. While other interferometers may differ in details, this should provide a reasonably complete description applicable to all of them.

1.4.3 The Palomar Testbed Interferometer

Figure 1.11: A picture of PTI taken from the catwalk of the Palomar 200-inch telescope.

The Palomar Testbed Interferometer (PTI) is a long-baseline infrared interferometer installed at Palomar Observatory, California. It operates in the J (1.2 μm), H (1.6 μm) and K (2.2 μm) bands, and with a maximum baseline of 110 m achieves an angular resolution of $\sim$ 3 mas. It was developed by the Jet Propulsion Laboratory, California Institute of Technology for NASA, as a testbed for interferometric techniques applicable to the Keck Interferometer as well as other missions such as the Space Interferometry Mission, SIM. PTI has been used in the development of high-sensitivity direct-detection interferometry in the infrared with array detectors, phase-referencing, and narrow-angle astrometry. PTI also serves as a testbed for interferometric planning, operational techniques, and data processing and management tools applicable to both ground and space-based interferometers.

Major development of PTI began in November 1992 with the commencement of funding from NASA under its TOPS program. The interferometer was installed at Palomar Observatory during the spring of 1995, and first fringes were obtained in July 1995. Initial

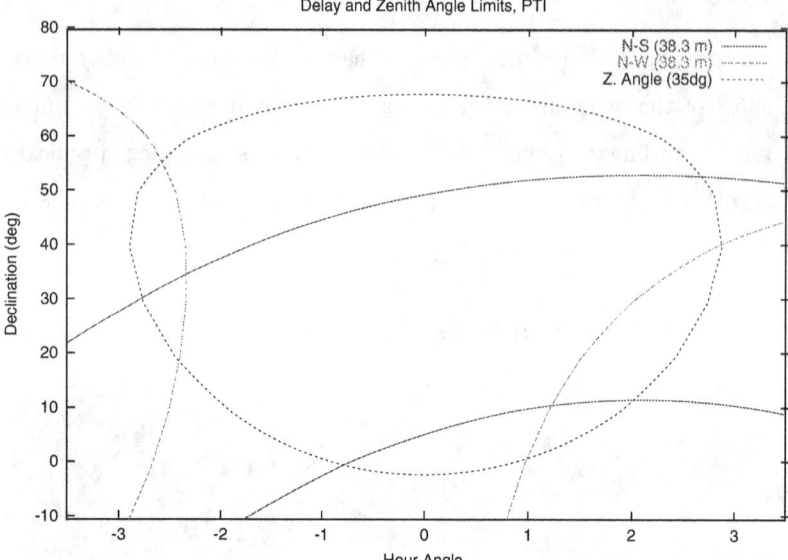

Figure 1.12: The sky coverage of the two PTI baselines, and the limit imposed by requiring the zenith angle to be less than 35 degrees, required for optimal fringe tracking performance.

astrometric measurements were made starting in 1997, achieving 100μas accuracy in 1999. Phase referencing results using long integration times were also first obtained in 1999.

1.4.3.1 Apertures, Baselines and Light Pipes

An interferometer requires two or more sub-apertures arranged so as to best sample the uv plane. In order for the light to have a high degree of coherence when combined, it is necessary that the wavefront sampled by each aperture be as uniform as possible; until recently this has meant that the interferometric sub-apertures could be, to within a factor of ~ 3, no larger than the atmospheric coherence length, r_0. Hence the current generation of interferometers all have aperture sizes of 10–50 cm[e].

Recently the Keck Interferometer and VLTI systems have been designed to make use of large (10 and 8 m respectively) apertures. This is possible because of the use of adaptive optics (AO) systems that operate in the near-IR (JHK bands, or 1–2.5 μm.), which typically bring $\sim 50\%$ of the light into a coherent or diffraction-limited beam. However, it should be

---

[e]One exception to this practice is the GI2T, which used 1.5 m apertures. However, it did not take full advantage of the large apertures, but instead operated in a multi-speckle mode.

noted that in the case of a natural guide-star based AO this does not necessarily increase the limiting magnitude of an interferometer; given the same photon throughput a star bright enough to use for an AO system is also bright enough to use in an interferometer.

The choice of aperture locations and hence baselines is of fundamental importance in designing an interferometer, and depends on the type of observations one wishes to perform. In the case of imaging, the goal is maximum coverage of the uv-plane, while for astrometry it is usually sufficient to have two perpendicular, long baselines. However, as the fringe visibility (and hence SNR of the measurement) drops for resolved objects (i.e., if $\theta > \lambda/B$), it is not desirable to only have very long baselines. These two competing requirements must be balanced in deciding the aperture placements. Typical imaging interferometers (COAST, NPOI) are laid out in a Y-pattern, with the apertures spaced evenly along the arms (as opposed to having VLA-style increasing separations); this provides acceptable uv coverage, while ensuring that there is always a short baseline available that can be used for fringe tracking and co-phasing. Such an approach to phasing an array is known as baseline bootstrapping (Hajian et al. 1999), and was originally developed for radio interferometry (Schwab & Cotton, 1983).

Once the incoming starlight has been collected in the sub-apertures, it must be directed to a central location where the beams can be combined and the fringe patterns measured. This is usually done by collimating the beams and directing them using mirrors. Although there would be advantages to using optical fiber to bring the light to the beam combiner, the severe practical difficulties associated with wavelength dispersion and polarization control have so far prevented such an approach[f]. Instead the light is "piped", and in the case of visible-wavelength systems, the pipes must be evacuated in order to avoid problems with dispersion. In order to minimize polarization effects the systems are laid out in such a way as to maintain optical symmetry.

PTI has three 40 cm apertures arranged in a triangle, although it presently can only combine light from two apertures at a time. The available baselines are 110 m (North-South) and 80 m (North-West). Although PTI only operates in the near-IR, and so does not strictly require evacuated light pipes, we have found that the internal metrology systems perform much better with an evacuated system; hence the beam pipes at PTI are held to a

---

[f]Note that optical fibers have been used in beam *combination* (FLUOR), but to date they haven't been used for beam *transport*.

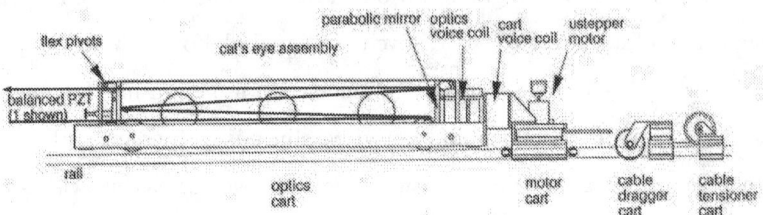

Figure 1.13: The PTI delay line cart. A cats' eye assembly (a parabola + flat mirror at the focus) is flexure-mounted on a cart that can move along the rails, propelled by the motor cart. To reduce vibrations the only connection between the motor cart and the mirror cart is via a voice-coil. The length of the rails is 19 m, giving a total delay range of 38 m. At the focus of the cats eye a small mirror is mounted on a PZT stage capable of rapid, short range motion, used in fringe tracking. The path taken by the starlight is indicated with red arrows. Figure from Colavita et al. (1991).

soft vacuum (~ 1 torr).

1.4.3.2 Acquisition & Tracking

The fact that the incoming starlight must be reflected down a small pipe over distances of tens to hundreds of meters in order to reach the beam combiner results in the systems having a very small field of regard (typically a few arcseconds). Hence most interferometers are equipped with a multi-stage acquisition and tracking system: the first stage is a wide field of view (arcminutes) imager on the telescope used to fine-tune the telescope pointing. Once the pointing is good enough that starlight is entering the central beam combiner, a fast (~ 100 Hz) tip-tilt star tracker is used to keep the star centered on the detectors, effectively correcting both internal vibrations and the first-order atmospheric seeing.

PTI has a R-band (0.5–0.7 μm) wide acquisition system based on a commercially available CCD camera. The PTI star tracker is based on a quad-cell feeding 4 avalanche photodiodes (APD's). It operates at a rate of 100 Hz, in the I band (0.8–1.0 μm).

1.4.3.3 Delay Lines & Metrology

Before being combined the incoming beams must be adjusted so that the total path-lengths (from the star, through the atmosphere and arm of the interferometer) are equalized to a coherence length or better (Eqn. 1.18 *et seq.*). This is done by delaying the light in each arm by different amounts, such that the total internal delay difference equals the external delay from geometry and the atmosphere. Note that the amount of required delay changes with sidereal motion, typical rates are a few mms^{-1}. In practice the delay is introduced by means of a mirror and cart assembly mounted on rails, capable of being controlled to small fraction of the operating wavelength while moving over distances of tens of meters.

The mirror assembly is typically either a "cats eye" or a corner-cube so as to maintain proper beam pointing despite imperfect rails, and the motion is controlled using a multi-stage servo system. Large-scale slow movement is provided by a motor pushing the cart, while intermediate range and rate motion is handled by flexure mounts and voice-coil actuators. In some cases very fine (tens of μm) motion at kHz rates is provided by a piezo-electric actuator pushing a small mirror (i.e., at the focus of the cats eye).

The path-length requirements on a delay line can be quite severe (at PTI rms path-length error should be < 20 nm), requiring a high-precision metrology system. This is usually accomplished using a dedicated laser metrology gauge; in effect a second Michelson interferometer. PTI uses a system based on a 633 nm HeNe single-mode laser.

1.4.3.4   Beam Combination & Detectors

The stabilized, delay compensated beams must be combined and the fringes measured using one of several possible beam combination schemes. Broadly, the beam combination methods can be characterized as occurring in either the image plane or the pupil plane. As implied by the name, an image plane combiner forms a fringe pattern in the image plane (i.e., identical to Fig. 1.5), which is measured with a position-sensitive detector such as a CCD array. By contrast, a pupil plane combiner combines the two (or more) beams at a beamsplitter[g], producing two output beams each containing half the light from each aperture. In this case the fringe pattern is detected by measuring the intensity in the combined beam as the relative delay is swept through one or more wavelengths; the detector used depends on the operating wavelength. A photo-multiplier or avalanche photo-diode is used for optical wavelengths, while single or multi-pixel InSb or HgCdTe detectors are usually used for IR operation. The pupil plane combiner has to date been the preferred approach, primarily because it is easier to implement a single-pixel detector. Due to the

---

[g]In some cases the beamsplitter is replaced with a fiber coupler, to the same effect (the FLUOR instrument).

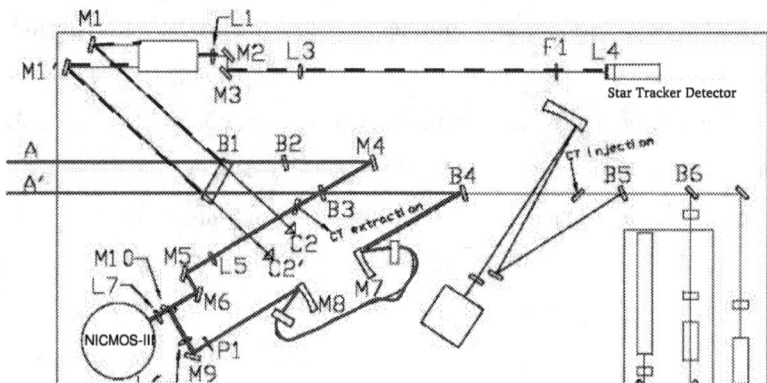

Figure 1.14: The PTI star tracker and beam combiner table. Near-IR (JHK) light enters from the left (A from North aperture, A' from the South/West aperture), and is combined at a 50/50 beamsplitter (B3). One combined output is re-imaged through L5–L7 onto a single pixel of a NICMOS-III infrared detector array; this output is used for real-time fringe tracking. The other combined output is directed via B4 into a single-mode fiber for spatial filtering, then dispersed by prism P1, and finally re-imaged onto 5–10 pixels of the same detector. This output is used for high-precision, spectrally resolved fringe visibility science measurements. The path taken by the JHK light used in the interferometric combiner is indicated with (solid) red arrows, while the I-band light used in the star tracker is reflected at beamsplitter B1. Its path is indicated with (dashed) blue lines. Beamsplitter B4 is 90% reflective and hence 10 % transmissive. It is used to inject artificial starlight for testing purposes, as well as laser metrology for path-length monitoring, and a boresight laser for alignment (via B5 & B6).

fringe motion introduced by the atmosphere, the fringe must be measured in a very short time, typically a few ms. Spectral information can be obtained by dispersing the beams, usually with a prism although grating and Fabry-perot filter designs have on occasion been used. Note that the short exposure time implies that it is not usually possible to achieve high dispersion and high SNR. PTI uses a 5-pixel spectrometer[h].

In order to improve the noise properties of the measured fringe visibilities, a spatial filter is usually employed. Such a device (either a pinhole at an intermediate focus, or a short length of single-mode optical fiber) only passes a single spatial mode of light to the beam combiner, rejecting all other modes which would otherwise combine incoherently on the detector. The effect is to bring the measurement precision in fringe visibility (V^2) from $\sim 10\%$ to $\sim 1\%$. At PTI a single-mode fiber is placed after the combination, and filters the spectral output.

1.4.3.5 Operation, Control & Data Processing

Most of the active systems outlined above requires some form of computer control and data recording system, typically implemented using microprocessors. At PTI the control system consists of 7[i] single-board[j] computers [k] running the VxWorks real-time operating system, with control software implemented in the C programming language. Operator control and data recording is via an integrated GUI running on a UNIX work station.

In normal operation an estimate of the fringe amplitude is produced every 10 or 20 ms; typically these estimates are averaged together over a 130-second integration time (a "scan"). Observations of a science target are interleaved with observations of 1–3 calibration stars repeatedly throughout the night. Routine observations with PTI are highly automated and follow a standard sequence: first pointing and target acquisition, then calibration measurements (of the stellar intensity "off the fringe", i.e., with the optical path difference deliberately displaced from zero), followed by fringe acquisition and tracking, and finally the apertures are pointed to dark sky and a dark-level calibration is obtained. The entire sequence takes approximately 5 minutes, and in a typical night PTI will produce

---

[h]$\lambda/\Delta\lambda = 12$

[i]One each for: 2 Fringe Trackers, Delay Line, Star Tracker, Acquisition, Siderostat and instrument Sequencer.

[j]VME Rack Mounted.

[k]Motorola 68040 processors.

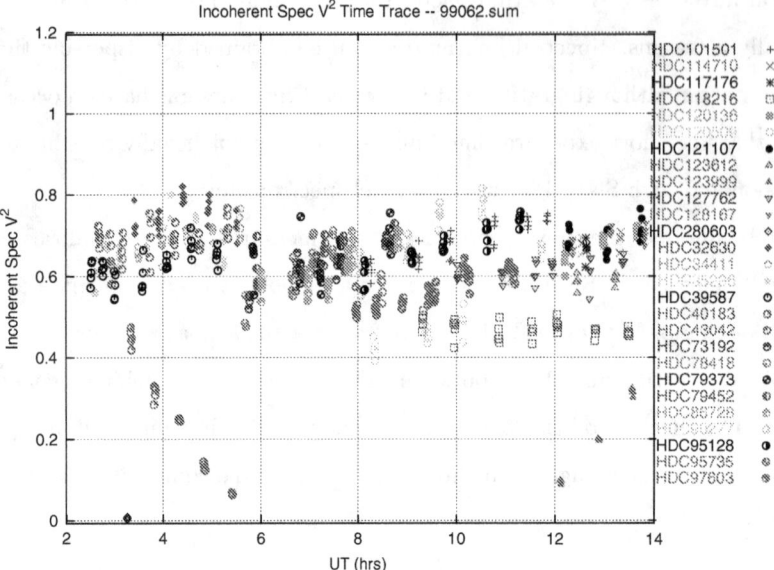

Figure 1.15: A night of PTI data (03 March 1999 = 99062). Observations of target and calibration sources are interleaved scan by scan. Each data point represents the 25-second average fringe visibility squared measured in the PTI spectrometer. Large variations in fringe visibility (HD 40183) typically indicate binarity, while the small-scale fringe visibility changes are related to instrumental effects and must be calibrated out by observing known single, unresolved stars.

60–120 scans.

During normal operation the instrument produces relatively large (300−800 Mbyte/night depending on mode) amounts of data. This also includes calibration data taken throughout the night, e.g. detector dark levels and relative intensities in the two arms of the interferometer. At the end of each night of observing all this data is processed to produce ∼ 100 Kbyte of "level-0" calibrated data. This calibrated set contains visibility, source brightness and delay-line metrology information and is used in later data analysis.

1.4.4 Differences between Optical/IR and Radio Interferometers

The basic principle of interferometry is equally applicable at any wavelength. However, there are a number of differences between a radio interferometer and an optical/IR interferometer. Most importantly, while most optical/IR interferometers (ISI excepted) use a direct-combination scheme outlined in section 1.4.3.4, most radio systems use *heterodyne detection*. In a heterodyne system, at each aperture the incoming electric field (frequency ν_s) is mixed with a local oscillator of known frequency (ν_{LO}) before being recorded. The mixing produces a harmonic $\nu_i = \nu_s - \nu_{LO}$ which can be low enough to be easily digitized and recorded. As a result the final beam combination can take place independently of the data collection (i.e., in VLBI). In addition, as the recorded field can be copied, it is possible to measure the fringes on a large number of baselines at the same time without loss of SNR; something that is not usually possible with a direct-detection scheme.

However, the presence of the local oscillator field introduces an unavoidable source of noise in the measurement, as follows from the uncertainty principle. The end result is that direct detection schemes enjoy an SNR advantage (assuming equal bandwidth and averaging time) of (Lawson, 2000)

$$\frac{SNR_{Direct}}{SNR_{Heterodyne}} = \sqrt{\frac{e^{h\nu/kT} - 1}{1 - \epsilon}} \qquad (1.57)$$

where ν is the observing frequency, T the system temperature and ϵ the system efficiency. For typical direct detection systems operating at room temperature, this implies an advantage factor of unity at ~ 120 μm, rising to ~ 17 at 10 μm and 10^5 at 2.2 μm. Hence the preference for direct detection.

1.5 Advanced Interferometric Techniques

Current state of practice in optical/IR interferometry provides a measurement precision of 1–10% in fringe visibility, which in turn limits the attainable dynamic range in any resulting images (or parametric models) to the same level. Clearly it would be desirable to improve the measurement precision. In addition, the current generation of interferometers is limited to observing comparatively bright sources (i.e., brighter than $\sim$ 8th mag); below I outline a number of techniques that are being developed to improve the performance of interferometers, in terms of both sensitivity and measurement precision.

1.5.1 Dual Star Interferometry

Most current interferometers have an extremely small interferometric field of view, on the order of 0.1 arcseconds. However, the field of view of the subapertures is only limited by the optical design, and can be much larger. It is therefore possible to split the image

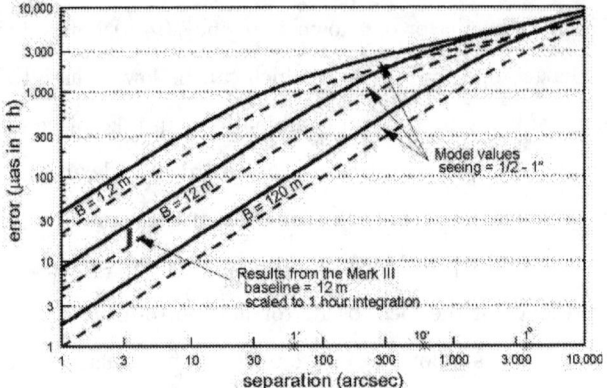

Figure 1.16: Astrometric accuracy vs. star separation in a one-hour integration for different baseline lengths. Model atmospheres providing 1/2- and 1.0-arcsecond seeing are shown. These results assume an infinite outer scale, and better results are achieved when the baseline exceeds the outer scale, as would be expected with a 100 m baseline at most sites. Measurements with the Mark III interferometer of a 3.3 arcsecondbinary star are consistent with the model. This figure is from Shao & Colavita, 1992.

plane of the subapertures into several "sub-fields" and interferometrically combine each subfield separately. This is called "dual star" interferometry, and is particularly useful for astrometry.

1.5.1.1   Astrometry

As discussed in Sec. 1.2, the primary observable of an interferometer is the complex visibility – a quantity which measures the degree to which the electric field at two positions are correlated. Complex visibilities are need to synthesize interferometric images, cf. radio interferometry. However, one can make use of an interferometer in a different way by taking advantage of a property of the visibility; as seen in Fig.1.6 the fringe pattern peaks at zero path-length difference. In a Michelson interferometer, the two apertures are fixed to the ground and the condition of zero path-length requires that we know the position of the source to a precision better than the resolution offered by the baseline ($\theta = \lambda/|\vec{B}|$). Conversely, the path-length inserted to obtain zero path-length yields the position of the source to high precision. Mathematically this is simply

$$d = \vec{B} \cdot \vec{S} + c \tag{1.58}$$

where d is the optical path-length one must introduce between the two arms of the interferometer to maximize the visibility. This quantity is often called the "delay." $\vec{B}$ is the baseline – the vector connecting the two apertures. $\vec{S}$ is the unit vector in the source direction, and c is a (hopefully) constant additional scalar delay introduced by the instrument.

Given a measured value for d (i.e., the optical delay necessary to drive the fringe phase to zero), and knowledge of $\vec{B}$ and c, one can invert the above equation to obtain $\vec{S}$. Unfortunately this is where the atmosphere causes problems, by introducing path-length fluctuations. The magnitude of these fluctuations determines the precision of the astrometric measurement; for a simple delay tracking optical interferometer the highest precision achieved is of order 5 mas – only a factor of a few worse than that of Hipparcos (Hummel et al. 1994, Armstrong et al. 1998, ESA 1997).

However, the limitations discussed above can be overcome by taking advantage of the fact that the path-length fluctuations introduced by the atmosphere are correlated over small angles on the sky (isoplanatism, see Sec.1.4.1). If we observe two stars separated by less than the isokinetic angle, we can (by definition) expect the atmospheric path-length fluctuations to be correlated, and to zeroth order they should subtract out. The isokinetic angle is given by

$$\theta_I = B/H \tag{1.59}$$

where B is the interferometer baseline, and H is the effective turbulence altitude.

With some effort it can be shown that the narrow-angle astrometric precision attainable is greatly improved (see Fig. 1.16). Reasonable calculations indicate that we can expect precision on the order of 10s of μas, and this has been experimentally verified (under very limited circumstances) at both PTI and the Mark III (Colavita 1994). For reference, an astrometric precision of 100 μas corresponds to knowing δ_d to 0.05 microns, a difficult but not impossible proposition. Specifically Shao and Colavita (1992) show that in the case of $\theta < \theta_I$, the variance of a narrow angle astrometric measurement is

$$\sigma_\delta^2 \simeq \frac{\theta^2}{tB^{4/3}} \int dh C_n^2(h) h^2 V^{-1}(h) \tag{1.60}$$

where t is the integration time, θ the angle between the two stars and $V(h)$ is the wind speed profile. For a typical Mauna Kea seeing profile this gives

$$\sigma_\delta \simeq 300 \frac{\theta}{\sqrt{t}B^{2/3}} \text{ arcsec} \qquad (1.61)$$

which for typical baselines of ~ 100 m, and an angular separation of ~ 30 arcsecond implies an astrometric precision of 30 μarcsec/$\sqrt{\text{hr}}$. Note that the astrometric precision is a strong function of turbulence height, and considerably better performance may be possible at sites lacking such high altitude turbulence (i.e., the South Pole, Lloyd, Oppenheimer & Graham 2002).

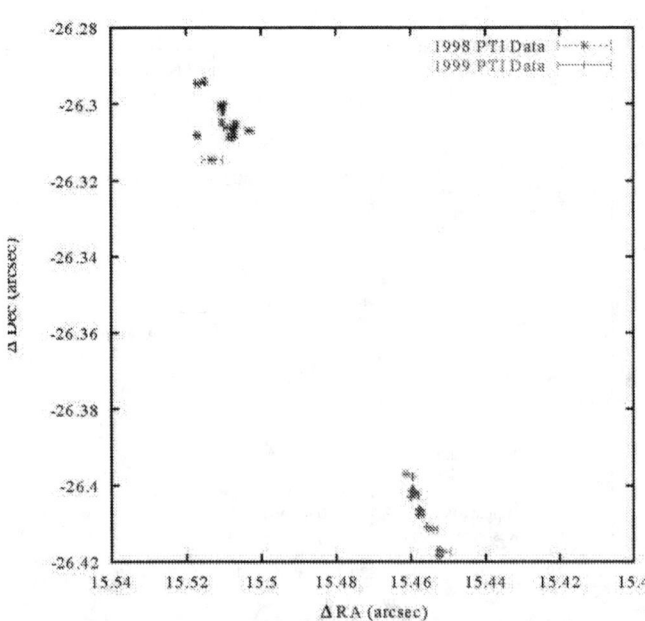

Figure 1.17: The differential astrometry between the components of 61 Cyg, as measured by PTI in 1998 and 1999. The numerous system and procedural improvements resulting in significantly better performance in 1999. Courtesy of A. Boden.

Astrometry at PTI During the summer of 1999 we observed the bright visual binary 61 Cygni (HD 201091/201092, K5V+K7V) to determine the astrometric performance of PTI. After numerous hardware improvements implemented as a result of our 1998 experience, 1999 data on 61 Cyg exhibits astrometric precision of 100 μas (10^{-6} arcseconds) over a one week timescale, and 170 μas rms precision over a 70 day timescale. Further, the 61 Cyg component differential proper motion measured by PTI in 1999 is in good agreement with the Hipparcos determination of system motion (Fig. 1.17).

However, at this point we have made the sensitivity problem worse – we now need two bright stars in close proximity on the sky. Such pairs do exist, usually in the form of nearby visual binaries such as 61 Cygni, but they are not sufficiently common to justify a large-scale planet search. This is where phase referencing fits in.

1.5.1.2 Phase Referencing

If one of the stars in the close pair is bright enough to track, the atmospheric variations can be monitored in real time and removed for *both* stars. This technique is referred to

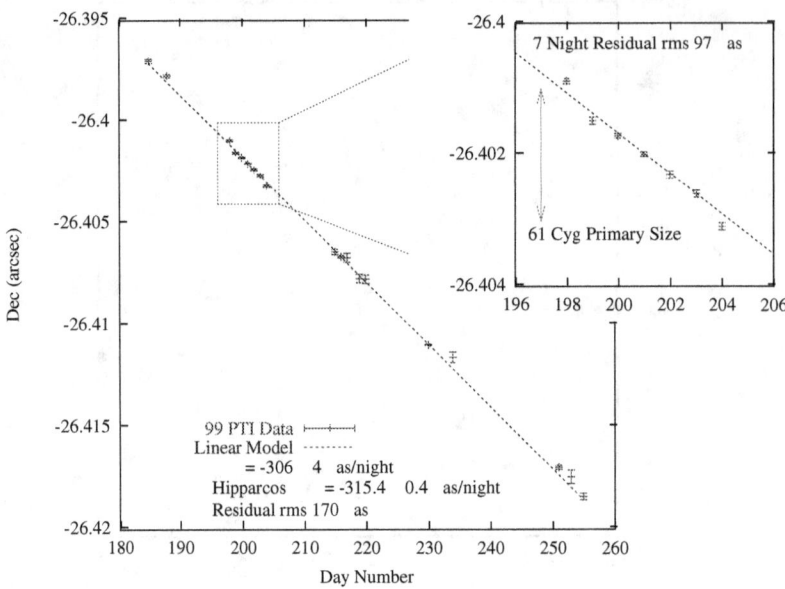

Figure 1.18: The differential declination between the components of 61 Cyg, as measured by PTI in 1999. Courtesy of A. Boden.

as "phase referencing." Phase referencing effectively increases the atmospheric coherence time (τ_0) as seen by the second tracker, allowing it to use longer integration times, with a correspondingly fainter tracking limit.

The ultimate achievable sensitivity gain from phase referencing is not well known. The gain depends on the properties of the atmosphere, the angular separation of the two stars, and the instrument stability. However, results at PTI show that an increase in exposure time by a factor of 25 (to 250 ms) is possible, see Chapter 5.

Thus our problem has changed – we need two stars in close proximity on the sky, but now one of them can be at least several (≈ 4) magnitudes fainter than the other. Although we still need one bright star to track, remember that all targets we would wish to study for planet-induced astrometric motion will necessarily be relatively close (in order for the astrometric signature which is inversely proportional to distance, to be large), and thus bright. The details of how this is done is discussed in Chapter 5.

1.5.2 Phase Closure

Despite the phase corruption introduced by the atmosphere, it is still possible to recover limited phase information without resorting to the technically complex phase referencing method, provided one interferometrically combines at least three apertures. In such a case one can form a quantity by multiplying the three complex visibilities formed over the three baselines; such a quantity is known as the "closure phase." The closure phase is immune to atmospheric corruption, which can be illustrated as follows: above each aperture there is a column of atmosphere with time-variable parcels of differing indices of refraction and hence optical pathlength. Thus the atmosphere above each aperture contributes a time-variable phase error, giving

$$\bar{V} = |\hat{V}|e^{i(\phi_{12}+\phi_1-\phi_2)} \tag{1.62}$$

where ϕ_1 and ϕ_2 are the phase errors associated with apertures 1 and 2, respectively, and ϕ_{12} is an intrinsic phase associated with the source as measured by the 1-2 baseline.

The quantity formed by multiplying the three complex visibilities corresponding to the three baselines is called the bispectrum, and is thus

$$
\begin{aligned}
\bar{V}_{123} &= |\hat{V}_1||\hat{V}_2||\hat{V}_3|e^{i(\phi_{12}+\phi_{23}+\phi_{31}+(\phi_1-\phi_2)+(\phi_2-\phi_3)+(\phi_3-\phi_1))} \\
&= |\hat{V}_1||\hat{V}_2||\hat{V}_3|e^{i(\phi_{12}+\phi_{23}+\phi_{31})}
\end{aligned}
\tag{1.63}
$$

We see that the atmospheric phase errors (as well as any other aperture-dependent phase errors) cancel. This is a well-known result, first applied in radio interferometry . The phase of the bispectrum is called the closure phase, and it depends only on the source and geometry of the observation. However, it is not immediately obvious what the closure phase represents. In general imaging applications, the closure phase is used as a constraint on the reconstructed map. In addition, since the number of independent closure phases increases rapidly with the number of apertures it becomes possible to recover more of the image phase information. However, for small numbers of apertures, the closure phase can be directly related to the image. Below we derive an expression relating the observed closure phase to the binary point source representing a binary star system.

Assume 3 apertures, resulting in 3 baselines $\vec{B}_1, \vec{B}_2$ and $\vec{B}_3$. Note that

$$\vec{B}_1 + \vec{B}_2 + \vec{B}_3 = 0. \tag{1.64}$$

As before, we are looking at two point sources with intensity ratio R and separation $\Delta \vec{s}$. On each baseline we measure a visibility $\hat{V}_n$ given by Eqn. 1.24.

$$
\begin{aligned}
V_{123} &= \hat{V}_1 \hat{V}_2 \hat{V}_3 \\
&= \frac{1}{(1+R)^3} \left(1 + Re^{-\frac{2\pi i}{\lambda} \Delta \vec{s} \cdot \vec{B}_1}\right) \left(1 + Re^{-\frac{2\pi i}{\lambda} \Delta \vec{s} \cdot \vec{B}_2}\right) \left(1 + Re^{-\frac{2\pi i}{\lambda} \Delta \vec{s} \cdot \vec{B}_3}\right) \\
&= \frac{1}{(1+R)^3} [1 + R(e^{-\frac{2\pi i}{\lambda} \Delta \vec{s} \cdot \vec{B}_1} + e^{-\frac{2\pi i}{\lambda} \Delta \vec{s} \cdot \vec{B}_2} + e^{-\frac{2\pi i}{\lambda} \Delta \vec{s} \cdot \vec{B}_3}) \\
&\quad + R^2 (e^{\frac{2\pi i}{\lambda} \Delta \vec{s} \cdot \vec{B}_1} + e^{\frac{2\pi i}{\lambda} \Delta \vec{s} \cdot \vec{B}_2} + e^{\frac{2\pi i}{\lambda} \Delta \vec{s} \cdot \vec{B}_3}) + R^3]
\end{aligned}
\tag{1.65}
$$

the closure phase is thus

$$\phi_{123} = \arctan\left(\frac{(R^2 - R) \sum_{i=1,2,3} \sin(\frac{2\pi}{\lambda} \Delta \vec{s} \cdot \vec{B}_i)}{1 + R^3 + (R + R^2) \sum_{i=1,2,3} \cos(\frac{2\pi}{\lambda} \Delta \vec{s} \cdot \vec{B}_i)}\right) \tag{1.66}$$

There are a few things to note: the closure phase is always zero when $R = 1$, i.e., the source is symmetric. In addition, by Taylor expanding the sine terms and recalling Eqn. 1.64 it is easy to show that for the case when $\Delta \vec{s} \ll \frac{\lambda}{|\vec{B}|}$

$$\phi_{123} \propto \left(\frac{\Delta \vec{s}}{\lambda/|B|}\right)^3 \tag{1.67}$$

This implies that a source must be resolved by the interferometer in order to produce a nonzero closure phase; in the case of a partially resolved source the magnitude of the closure phase signal is very sensitive to the separation of the source components.

Unlike V^2, the phase measured by an optical interferometer is largely unbiased by measurement noise. In other words, the phase noise is zero-mean, and can be reduced by averaging over a sufficiently long time. The SNR for closure phase is given by (Shao & Colavita 1992a)

$$\text{SNR}_\phi = \left[3\left(\frac{2}{NV^2}\right) + 6\left(\frac{2}{NV^2}\right)^2 + 4\left(\frac{2}{NV^2}\right)^3\right]^{-1/2} \tag{1.68}$$

Aperture (m)	Wavelength (μm)		
	1.2	2.2	5
1.8	10	43	300
10	0.3	2	10

Table 1.1: Required integration times to reach a closure phase measurement uncertainty of 10^{-4} radians, in seconds, for a 5th magnitude star. I have assumed a system throughput of 10%.

as compared to that for visibility and phase,

$$\text{SNR}_V = \left[\left(\frac{1}{NV^2}\right) + \left(\frac{1}{NV^2}\right)^2\right]^{-1/2} \tag{1.69}$$

In the photon-noise dominated regime ($NV^2 \gg 1$), the SNR for closure phase and visibility scales as $N^{1/2}$. However, for photon-starved sources, the SNR drops precipitously as $N^{3/2}$, even worse than the $\propto N$ scaling of the visibility SNR. Hence it is important to check whether sources will be photon-rich or photon-starved. The required integration time to achieve $\sigma_\phi \sim 10^{-4}$ radians, as a function of wavelength and aperture is shown in Table 1.

It is clear that the required SNR to measure a closure phase to the desired accuracy of 10^{-4} radians can be achieved with reasonable apertures and integration times. Note that since the atmospheric phase effects have disappeared, the coherence time (and thus maximum integration time) is now a function of the instrumental stability, which can be much longer (minutes to hours, depending on the thermal stability of the interferometer).

To date, closure phase measurements have been made in the optical and near-IR by 2 groups (COAST and NPOI). NPOI achieves phase drifts of ~ 10 degrees hr^{-1}, which can be calibrated to the level of 1-4 degrees by looking at known single stars (Hajian et al. 1998). Thus it is clear that closure phase techniques are currently systematics-limited, and will require further development before their full potential can be realized. This is where differential techniques may play an important role.

1.5.3 Differential Interferometry

Differential interferometry makes use of the fact that for most interferometric observables, many error sources have known wavelength dependences, which differ from that of the source. The first example of differential interferometry is "differential phase." In this

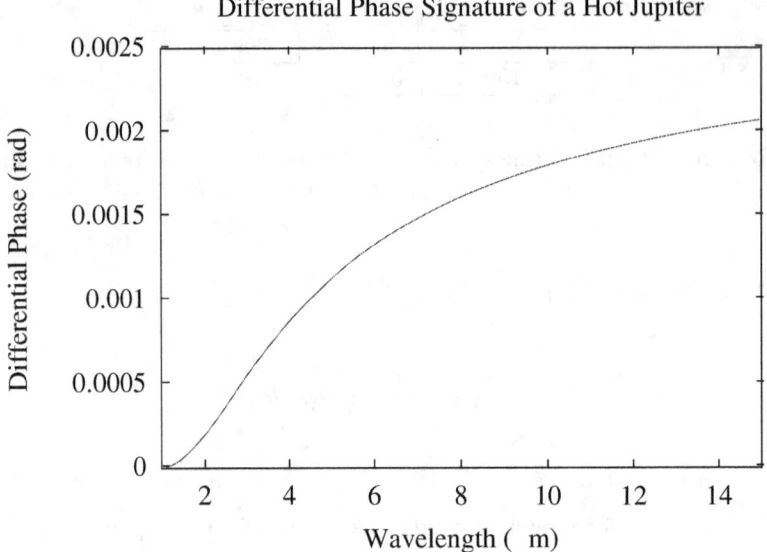

Figure 1.19: Calculated differential phase signature for a Hot Jupiter type planet, as a function of wavelength. The phase difference is calculated with respect to a reference wavelength of 1μm.

technique one observes a Hot Jupiter type system across a wide range of wavelengths simultaneously. Given that the star/planet intensity ratio changes drastically with wavelength (10^5 at 1μm to 10^3 at 3μm), the center-of-light (and hence fringe phase) should change with wavelength given by

$$\delta\phi = \phi(\lambda_1) - \phi(\lambda_2) \tag{1.70}$$

for small intensity ratios $\delta\phi \simeq R(\lambda_1) - R(\lambda_2)$, which is on the order of $\sim 10^{-4}$ radians (Fig. 1.19). The major advantage of a differential measurement is that the any phase effects introduced by the atmosphere *or in the instrument* are (to 1st order) common to both wavelengths. Thus they should cancel. Unfortunately, this cancellation is not perfect: although the two different wavelengths of light may take the same path through the instrument and atmosphere, the index of refraction of air varies with wavelength. This introduces a differential phase term, given by

$$(\Delta\phi)_{atm} = (n_{\lambda_2} - n_{\lambda_1})L \tag{1.71}$$

where L is the length of the path differential through the atmosphere.

Recently much work has been done by the Keck Interferometer (KI) design team to characterize the magnitude of the effects of dispersion on differential phase. At present the major source of concern is water vapor fluctuations. One lesson learned has been that the index of refraction of water vapor in the near-IR has is not well known. Measurements done at PTI indicate that a mere 76 μm of water vapor column will induce a differential phase signature of 2.85 radians across the K band.

One the best (driest) 20% of nights at Keck, the water vapor fluctuations in the atmosphere above the instrument have an RMS amplitude of 6 μm. Thus the atmosphere induces differential phase noise at the 0.2 radian level, on timescales of less than a minute. Clearly, in order to measure the differential phase signature of the planet, some method of measuring the water vapor fluctuations must be used. At present the KI team is working on a method of using multiple wavelengths to simultaneously solve for both the planet signature and the atmospheric effects.

A second differential technique is "differential closure phase", where one measures the closure phase across a range of wavelengths. It is similar to the differential phase technique in that it is well suited for observing Hot Jupiter-type systems. In addition, it does not suffer from the effects of dispersion, and so promises considerably better performance.

One important source of systematic errors in closure phase is thermal drifts in the optics within the beam combiner of the instrument. Although the closure relation causes all aperture-dependent phase errors (due both to the atmosphere and internal optics) to cancel, there are a few locations where baseline-dependent errors can be introduced. Such errors do not cancel, and must be accounted for. As an example, consider a case in which a baseline-dependent error of length δl is introduced in the path.

Consider as above a beam combiner that introduces a baseline-dependent error of length δl, thus biasing the closure phase. Now, if we measure the closure phase at two different wavelengths λ_1 and λ_2, what do we see? Our two closure phases are

$$
\begin{aligned}
\phi_1 &= \phi_{Source,1} + \frac{2\pi}{\lambda_1}\delta l \\
\phi_2 &= \phi_{Source,2} + \frac{2\pi}{\lambda_2}\delta l
\end{aligned}
\tag{1.72}
$$

Now convert the phases to distance

$$\begin{aligned}\frac{\lambda_1}{2\pi}\phi_1 &= \frac{\lambda_1}{2\pi}\phi_{Source,1} + \delta l \\ \frac{\lambda_2}{2\pi}\phi_2 &= \frac{\lambda_2}{2\pi}\phi_{Source,2} + \delta l\end{aligned} \qquad (1.73)$$

and subtract, forming the *differential closure phase* (DCP)

$$\begin{aligned}\left(\phi_1 - \frac{\lambda_2}{\lambda_1}\phi_2\right) &= \left(\phi_{Source,1} - \frac{\lambda_2}{\lambda_1}\phi_{Source,2}\right) \\ &\equiv \Delta\phi_{1,2}\end{aligned} \qquad (1.74)$$

Note that this quantity is now independent of δl. Thus we have found a way to also eliminate baseline-dependent errors. Note that because we are looking at the difference between two noisy quantities, the SNR is lower by a factor of $\sqrt{2}$. In addition, this technique requires at least three apertures with baseline separations sufficient to resolve the planetary system. However, we have now eliminated the two major error sources: the atmosphere and instrumental drifts. This technique is being studied for use at the VLTI.

1.5.4 Nulling

One of the primary observing modes planned for the Keck Interferometer is "nulling"; in this mode the light from a nearby star is combined such that the on-axis starlight is nulled out (destructive interference) while the off-axis light emitted by circumstellar material is detected. Because the exozodiacal emission is strongest in the thermal infrared ($\sim 10\ \mu$m) this is the wavelength region where the KI nuller will operate. As with all observations in that wavelength regime, the intense thermal background necessitates special measures. KI will make use of a novel interferometric chopping technique, as well as the well-known but still experimental interferometric nulling. Hence it is fair to say that KI will be very technically challenging. However, the KI nulling mode will also be a uniquely capable scientific instrument. The combination of very high-contrast imaging (by virtue of the null depth of 10^3 effectively removing the central star), and operation in the thermal IR makes KI able to not only detect exo-zodiacal dust emission, but also study circumstellar disks in young planetary systems, and even a few close-in massive planets.

1.6 Outline of Presentation

This thesis consists of two parts: a technical development effort aimed at demonstrating phase referencing for use in astrometry, and a scientific study of astrophysically interesting sources (Cepheids & low mass stars) using the high angular resolution measurements made possible by PTI. In addition I include adaptive optics-based high angular resolution observations of the very low mass binary GJ 569B; while not based on interferometry, these observations demonstrate the value of high angular resolution. In addition, adaptive optics shares many of the technical challenges of interferometry, and indeed a functioning AO system is a prerequisite for interferometry with large telescopes.

In Chapter 2, I describe how PTI can be used to obtain high precision angular diameter measurements of stars, and use this technique to empirically verify current mass-radius-luminosity relations for the lower main sequence. In Chapter 3, I use high precision angular diameter measurements to resolve the pulsation-induced diameter changes in two Galactic Cepheid variables. Together with previously published radial velocity measurements, these observations allow me to determine the distances to these fundamentally important stars. In Chapter 4, I use PTI fringe visibility measurements to resolve the low mass binary star BY Dra and hence obtain mass estimates for the stellar components of that system. Chapter 5 details development of the phase referencing technique, and initial observations made at PTI. Chapter 6 details the adaptive optics-based investigation of the low mass binary GJ 569B, which likely contains at least one substellar component. Appendix A is a review of basic control theory, and a derivation of the servo response functions of the PTI fringe tracker in various modes, including phase referencing.

The work described in this thesis has resulted in 6 refereed publications: Chapter 2 is based on Lane, Boden & Kulkarni (2001), Chapter 3 is based on Lane et al. (2000) and Lane, Creech-Eakman & Nordgren (2002). Chapter 4 is based on work done in Boden & Lane (2000), Chapter 5 is currently in press as Lane & Colavita (2003), and Chapter 6 has been published as Lane et al. (2001).

Chapter 2

Stellar Diameter Measurements

We have used the Palomar Testbed Interferometer to measure the angular diameter of 5 dwarfs of spectral types K3-M4. Using the 110-meter baseline and observing in H and K bands allows us to achieve a measurement accuracy of 2–8% on stars with apparent angular diameters approaching 1 milli-arcsecond. We provide results for both uniform disk and limb-darkened models, and compare our results with theoretical predictions. At the current level of precision our measurements are consistent with most widely accepted models, but further observations should be able to provide useful empirical constraints.

2.1 Introduction

The essential link connecting conventional observational data (colors, spectra, photometry) of stellar populations to physical models of the stars (structure, evolution, atmospheres) is the empirical determination of how fundamental stellar properties such as radius and luminosity depend on mass. There are only a few ways to determine in a model-independent way these properties: for radius determination, observations of eclipsing binaries, lunar occultation, and interferometry are the only available methods. These methods have resulted in relatively accurate calibrations (2–5%, enough to constrain models) for most early-type stars (Andersen 1991).

However, for the lower main sequence the data is much more sparse, despite the fact that such stars dominate the stellar census. In particular, there are only three M-dwarf systems for which model-independent mass-radius-luminosity determinations have been made: YY

---

[a]The material in this chaper was previously published as Lane, B., Boden, A.,& Kulkarni, S. R., ApJL, 551, L81-L83, 2000.

Gem, CM Draconis and GJ 2065A (Leung & Schneider 1978, Metcalfe et al. 1996, Delfosse et al. 1999). Although there has been much progress in modeling the atmospheres of these late-type stars (Charbrier et al. 1995, Allard et al. 1997) independent observational constraints are still desirable.

This lack of precision measurements makes it difficult to assess the contribution of M dwarfs to the total mass of the Galaxy. The exact behavior of the mass-radius relation in this regime may also be of interest for other reasons, as Clemens et al. (1998) claim that the mass-radius relation steepens between 0.2 and 0.3 solar masses and this steepening is the cause of the well known gap in orbital periods of cataclysmic variables.

The MLR relation needs to be defined empirically since the physics of M dwarfs is quite complicated (Chabrier & Baraffe 1995, Allard et al. 1997). Not only molecules with their numerous transitions but also dust starts contributing to the opacity below 3000 K (Jones & Tsuji 1997). Even their supposedly simpler, fully convective interiors may be complicated. For example, Clemens et al. (1998) question whether gradients in mean molecular weight (μ) may develop and also worry about nonideal corrections to the interior equation of state.

The importance of improving fundamental stellar parameters has not escaped the attention of astronomers. Henry et al. (2000) have proposed a program of accurate mass determination with the Space Interferometry Mission and Clemens et al. (1998) stress the importance of increasing the sample of eclipsing M dwarf binaries with the goal of measuring their radii. Long baseline interferometry offers a method by which the radii of the nearby M dwarfs can be measured. Here we report direct measurements of the apparent angular diameters of several nearby dwarf stars using the Palomar Testbed Interferometer (PTI). PTI is 110-m long single-baseline infrared direct-detection interferometer located on Palomar Mountain, California (Colavita et al. 1999).

2.2 Observations

We selected a small number of relatively bright dwarf stars in the spectral range K3-M4 (see Table 2.1) and observed them with PTI in order to determine their apparent angular diameters. Each object (science target and 2–3 calibrators) was observed for a 130-s integration 4–8 times per night, during at least two nights during the 1999 and 2000 observing seasons.

Calibrators were selected and observed so as to minimize both time- and position-dependent changes in the system response; observations of calibrators and science targets were interleaved over a 10-minute timescale, and calibrators were chosen so as to be no more than 10 degrees away from the science targets on the sky. Associated with each 130-s integration a 25-s measurement of the background light level was obtained by off-pointing the interferometer apertures. Observations were obtained in both H and K bands on separate nights. Further discussion of the data reduction and calibration procedures used are available in Colavita (1999b).

2.3 Visibility and Limb Darkening

The interferometric observable measured by PTI is the contrast or visibility of the fringes that are produced when starlight from two apertures is combined. For this work we assumed the intensity profile of the stars of interest to be well approximated by a linear limb-darkening law

$$I(\mu) = I(1)(1 - u_\lambda(1 - \mu)) \tag{2.1}$$

where μ is the cosine of the incidence angle and u_λ is the linear limb-darkening coefficient. We used passband-specific linear limb-darkening coefficients from Claret et al. (1995).

From basic interferometric theory it follows that the visibility of such a limb-darkened disk of angular diameter θ_{LD} is given by Hanbury-Brown et al. (1974):

$$V = \left(\frac{1 - u_\lambda}{2} + \frac{u_\lambda}{3}\right)^{-1} \left[\frac{(1 - u_\lambda)J_1(\pi B\theta_{LD}/\lambda)}{\pi B\theta_{LD}/\lambda} + \frac{u_\lambda(\pi/2)^{1/2}J_{3/2}(\pi B\theta_{LD}/\lambda)}{(\pi B\theta_{LD}/\lambda)^{3/2}}\right] \tag{2.2}$$

where B is the projected baseline length and λ is the observing wavelength. J_1 and $J_{3/2}$ are Bessel functions of the first and three-halves order, respectively. We define the uniform-disk diameter θ_{UD} as the diameter of a model in which $u_\lambda = 0$.

2.4 Calibration

In order to correct for the inherent loss of fringe visibility due to the instrument and especially the atmosphere, we used calibrator stars of known diameter to determine the response function of the instrument (called the *system visibility*). The measured visibility of the sci-

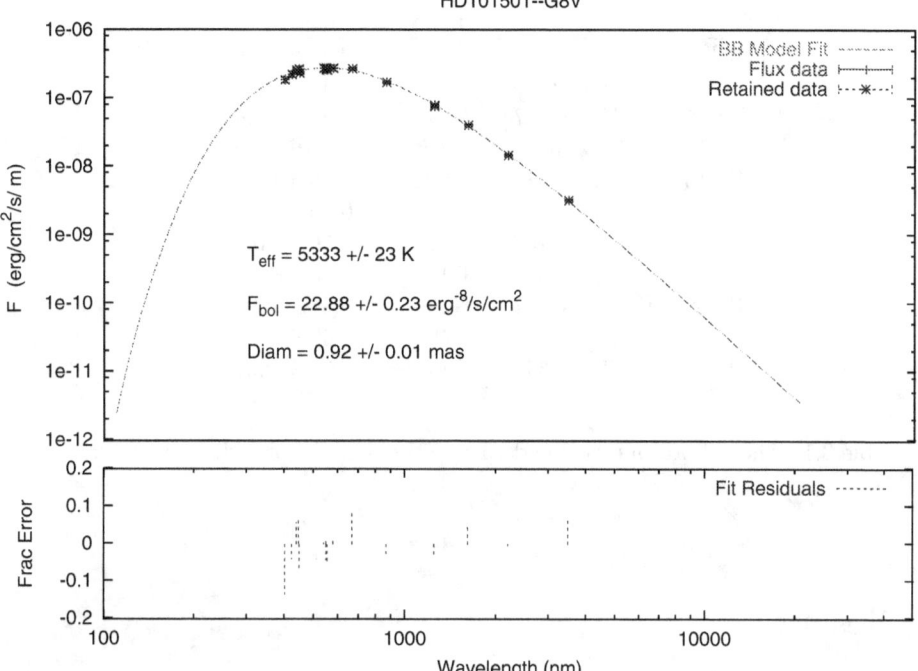

Figure 2.1: An example of the black-body fits used to estimated the diameters of the calibrator stars.

ence object was then calibrated by dividing it by the system visibility. Calibrators were selected so as to be as point-like as possible; thus even though the fractional uncertainty in diameter may be relatively large, because the apparent diameters of the calibrators are much smaller than the instrument resolution the resulting uncertainty in system visibility (and thus also diameter of the target star) is small. As an example, the 7% uncertainty in diameter of HD 171834 only produces a 0.7% uncertainty in the system visibility.

We determined the apparent diameters of the calibrator stars by using archival photometry to fit a blackbody model for the bolometric flux of the star in question, while either simultaneously fitting for the effective temperature of the star, or constraining it to the expected value based on the spectral type. We also calculated the expected diameter using the expected physical size based on spectral type (Allen 1982) and the Hipparcos (ESA 1997) distance to the star. We adopted the weighted mean of the results from all three methods as the final diameter, and the uncertainty in the determination was taken to be the deviation. The calibrators and their estimated sizes and uncertainties are given in Table

Primary Star	Calibrators	Spectral Type	θ_{UD} (mas)
HIP87937	HD 161868	A0V	0.64 ± 0.06
	HD 171834	F3V	0.42 ± 0.03
HD 95735	HD 90277	F0V	0.58 ± 0.06
	HD 101501	G8V	0.91 ± 0.02
HD 1326	HD 1671	F5III	0.65 ± 0.08
	HD 1279	B7III	0.19 ± 0.07
	HD 6920	F8V	0.565 ± 0.01
	HD 7034	F0V	0.467 ± 0.067
HD 88230	HD 84737	G2V	0.81 ± 0.005
	HD 89744	F7V	0.52 ± 0.02
HD 16160	HD 7034	F0V	0.36 ± 0.02

Table 2.1: The calibrators used and their estimated uniform-disk diameters.

2.1.

2.5   Results

Apparent angular diameters for the target stars were determined by fitting both uniform disk and limb-darkened models to the calibrated visibilities; the uniform disk model is provided, despite being less physically accurate, in order to allow follow-up work using different limb-darkening corrections. An example of the fits is shown in Figure 2.2 and results are given in Table 2.3. The differences between uniform-disk and limb-darkening models are too small to be readily apparent in the plot, but do amount to a few percent in diameter.

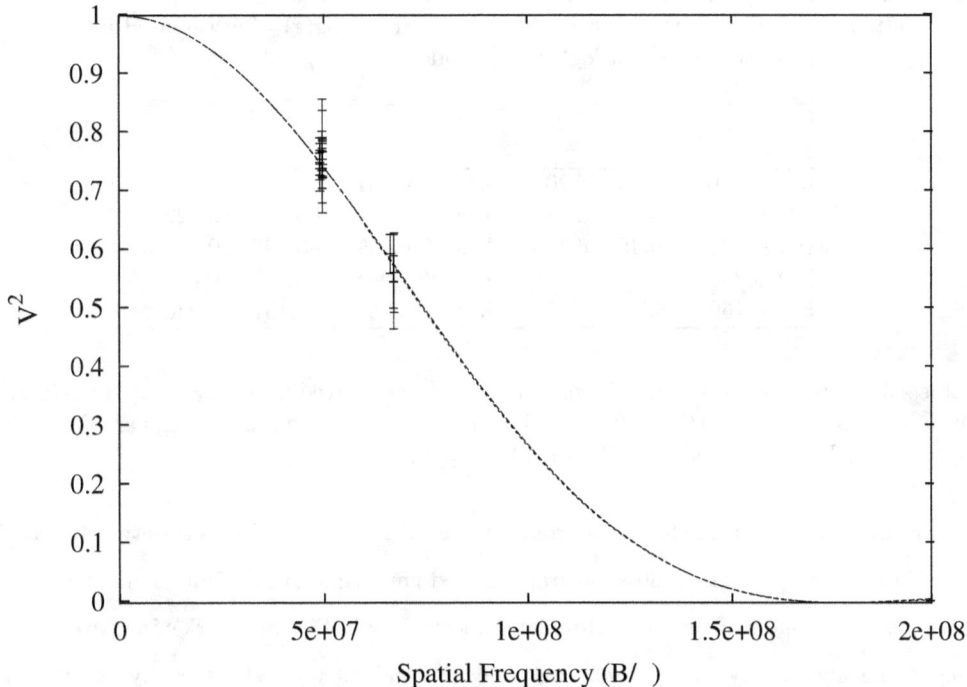

Figure 2.2: Measured visibilities (V^2) for HD 95735, together with the best-fit limb-darkened disk model. The model is given by Equation 2. The two clusters of data points correspond to observations in K and H bands.

Object	Spectral Type	u_λ 1.6μm	u_λ 2.2μm	Diameter θ_{UD}	Diameter θ_{LD}	Uncertainty σ_{Total} ($\sigma_{stat}/\sigma_{sys}$)	No. Scans	χ^2_{red}
HIP 87937	M4V	0.513	0.427	0.987	1.026	0.04 (0.013/0.035)	12	0.2
HD 1326	M2V	0.407	0.335	1.023	1.051	0.03 (0.029/0.006)	53	2.3
HD 95735	M1.5V	0.391	0.322	1.413	1.464	0.03 (0.026/0.015)	16	0.4
HD 88230	K7V	0.397	0.328	1.268	1.175	0.04 (0.042/0.005)	9	0.4
HD 16160	K3V	0.442	0.378	0.914	0.941	0.07 (0.027/0.064)	19	0.4

Table 2.2: Measured angular diameters for the target stars, for both uniform-disk and limb-darkened models. Also shown are published spectral types of the target stars, along with linear limb-darkening parameters from Claret et al. (1995), selected using the appropriate effective temperature and gravity; a model calculated for log(g)=4.5 was used in all cases except HD 87937 where we used a log(g)=5.0 model.

Object	M_V	M_K	Log($M/M_\odot$)	Log($R/R_\odot$)
HIP 87937	13.24	8.20	-0.833 ± 0.067	-0.697 ± 0.017
HD 1326	10.31	6.28	-0.387 ± 0.089	-0.395 ± 0.022
HD 95735	10.46	6.34	-0.394 ± 0.089	-0.397 ± 0.010
HD 88230	8.17	4.77	-0.184 ± 0.065	-0.211 ± 0.017
HD 16160	6.53	4.15	-0.113 ± 0.065	-0.137 ± 0.036

Table 2.3: Physical parameters. The mass estimate was derived from M_K using the relation given in Henry & McCarthy (1993). The radius measurement used angular diameters measured by PTI and parallaxes obtained by Hipparcos.

The uncertainties in the fits come from three sources: statistical uncertainty (estimated from the internal scatter in a 130-s integration), systematic uncertainty due to uncertainty in the angular diameters of the calibrators, and uncertainty in the limb-darkening parameters used. The uncertainty in the limb darkening parameters was estimated by adopting a range of parameters corresponding to an uncertainty in effective temperature of 200 K. These uncertainties were propagated separately in the fits, and added in quadrature to derive a total uncertainty. Finally, the observed limb-darkened angular diameters were converted into linear radii using parallax data from Hipparcos, and the associated parallax uncertainties were added in quadrature to the total uncertainty.

In Figures 2.3 and 2.4 we compare the measured diameters with the expected values based on a theoretical (Baraffe & Chabrier 1996) and an empirical model; the latter is the $[M_V, V - I]$ fit derived by Reid & Gizis (1997) transformed in the same manner as in Clemens et al. (1998) (who derive both log(R) and log(M) as a function of M_V). For Figure

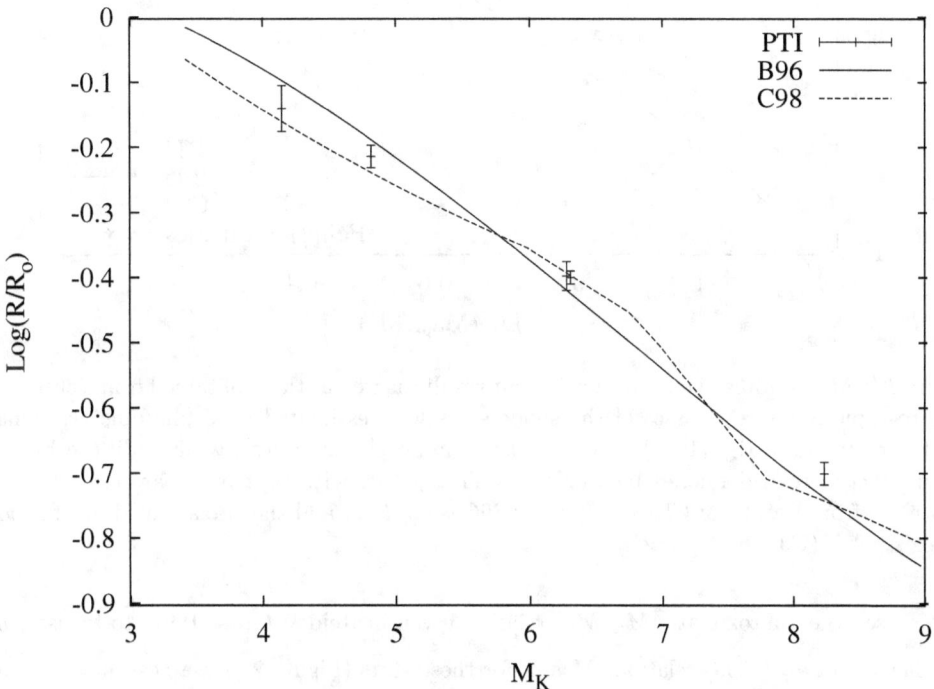

Figure 2.3: M_K-Radius diagram. Models shown are from Baraffe & Chabrier (1996) (B96, *solid line*) and Clemens et al. (1998) (C98, *dotted line*).

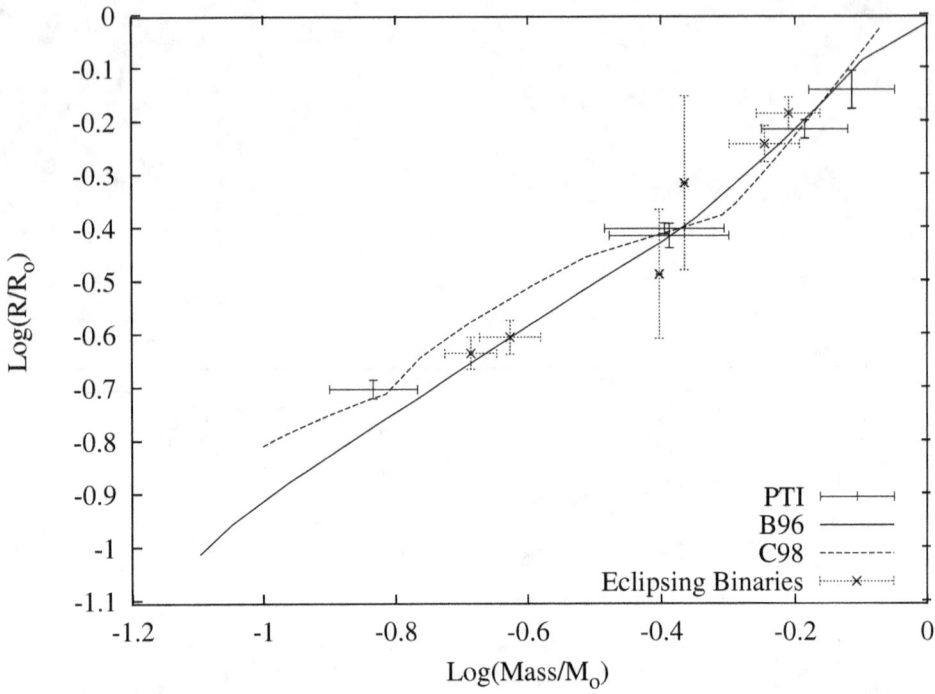

Figure 2.4: Mass-radius diagram, showing our results as well as those obtained from eclipsing spectroscopic binaries. Masses of the single stars were estimated from photometry, using the Henry & McCarthy (1993) relation (based on speckle binaries), while radii are based on apparent angular diameters measured at PTI, together with Hipparcos parallax. Models shown are from Baraffe & Chabrier (1996)(B96, *solid line*) and the transformed fit of Reid & Gizis (1997)(C98, *dotted line*).

2.3 we used the approximate $[M_K, M_V]$ relation given in Reid & Gizis (1997) to transform the Clemens et al. (1998) relation. Masses for these stars (Figure 2.4) were estimated from photometry using the mass-M_K relation from Henry & McCarthy (1993).

As can be seen in Figures 2.3 and 2.4, at the current level of measurement precision the measured diameters are consistent with the models. However, distinguishing between models will require further observations. While some of these observations can be obtained with PTI, longer baseline interferometers such as CHARA and NPOI will be able to provide many more such diameter measurements. In addition, the Keck Interferometer with its greater sensitivity should be able to contribute measurements of diameter of late M dwarfs including those close to the substellar boundary. The same interferometers in conjunction with newly commissioned IR spectrometers will also be useful in improving mass estimates of M dwarfs through observations of binaries.

Chapter 3

Cepheid Distances and Diameters

We present observations of the Galactic Cepheids η Aql and ζ Gem. Our observations are able to resolve the diameter changes associated with pulsation. This allows us to determine the distance to the Cepheids independent of photometric observations. We determine a distance to η Aql of 320 ± 32 pc, and a distance to ζ Gem of 362 ± 38 pc. These observations allow us to calibrate surface brightness relations for use in extra-Galactic distance determination. They also provide a measurement of the mean diameter of these Cepheids, which is useful in constructing structural models of this class of star.

3.1 Introduction

The class of pulsating stars known as Cepheids is a cornerstone in determining the distances to nearby galaxies. This is because Cepheids exhibit a well-behaved period-luminosity relation which can be locally calibrated (Jacoby et al. 1992). In addition, these stars are massive and thus intrinsically very luminous, making it possible to observe Cepheids located in very distant galaxies (Tanvir 1999, Feast & Catchpole 1999). Because of the usefulness and fundamental importance of Cepheids, it is important to calibrate their period-luminosity relation. This has been done using a variety of methods, including parallax (ESA 1997, Feast & Catchpole 1997), Baade-Wesselink methods (Wesselink 1946, Bersier et al. 1997) and surface brightness (Laney & Stobie 1995, Fouque & Gieren 1997, Ripepi et al. 1997). The period-luminosity relations used currently have uncertainties on the order of 0.09 mag (Feast 1999), which in turn make up a significant portion of the systematic uncertainty

---

[a]The material in this chaper was previously published as Lane, Creech-Eakman & Nordgren, ApJ, 573, 330–337, 2002.

in estimates to the Large Magellanic Cloud.

Using long-baseline stellar interferometry it is possible to resolve the diameter changes undergone by a nearby Cepheid during a pulsational cycle. When such diameter measurements are combined with radial velocity measurements of the stellar photosphere, it is possible to determine the size of and distance to the Cepheid. Such a direct measurement is independent of photometric observations and their associated uncertainties.

The Palomar Testbed Interferometer (PTI) is located on Palomar Mountain near San Diego, CA (Colavita et al. 1999). It combines starlight from two 40 cm apertures to measure the amplitude (a.k.a. visibility) of the resulting interference fringes. There are two available baselines, one 110-m baseline oriented roughly North-South (hereafter N-S), and one 85-m baseline oriented roughly North-Southwest (called N-W). In a previous paper (Lane et al. 2000), we presented observations using PTI of the Cepheid ζ Gem. Here we report on additional interferometric observations of ζ Gem, as well as a second Galactic Cepheid, η Aql. These observations allow us to determine the distances to these Cepheids with the aim of reducing the uncertainty in currently used period-luminosity relations for Cepheids.

3.2   Observations

We observed the nearby Galactic Cepheids η Aql and ζ Gem on 22 nights between March 13 and July 26, 2001. The observing procedure followed standard PTI practice (Boden et al. 1998, Colavita 1999) . For the observations of η Aql the N-W baseline was used, while observations of ζ Gem used the N-S baseline. Each nightly observation consisted of approximately ten 130-second integrations during which the fringe visibility was averaged. The measurements were done in the $1.52 - 1.74$ μm (effective central wavelength 1.65 μm) wavelength region, similar to the astronomical H band. Observations of calibration sources were rapidly (within less than $\sim$ 10 minutes) interleaved with the Cepheid observations, and after each 130-second integration the apertures were pointed to dark sky and a 30-second measurement of the background light level was made.

The calibrators were selected to be located no more than 16 degrees from the primary target on the sky and to have similar H-band magnitudes. In choosing calibration sources we avoided known binary or highly variable stars. The calibrators used are listed in Table

Star Name	Alternate Name	Period (d)	Epoch JD	Limb Dark. Factor (k)
η Aql	HD 187929	7.176711	2443368.962	0.97 ± 0.01
ζ Gem	HD 52973	10.150079	2444932.736	0.96 ± 0.01

Table 3.1: Relevant parameters of the Cepheids. The limb darkening factor is defined as $k = \theta_{UD}/\theta_{LD}$.

3.2. In this paper we make use of previously published observations of the Cepheid ζ Gem (Lane et al. 2000). However, in order to improve on the previously published results we performed additional observations of this source on March 13–15, 2001. We also observed additional unresolved calibrators in order to reduce the level of systematic uncertainty. The original data have been jointly re-reduced using the improved calibrator diameters and uncertainties. However, note that the primary calibrator diameter has not changed from the value used in Lane et al. (2000).

Calibrator	Spectral Type	Diameter Used θ_{UD} (mas)	Limb Dark. Factor (k)	Used to calibrate	Cal. Type	Angular Sep. (deg)
HD 189695	K5 III	1.89 ± 0.07	0.943 ± 0.007	η Aql	Pri. Cal	7.8
HD 188310	G9.5 IIIb	1.57 ± 0.08	0.955 ± 0.007	η Aql	Sec. Cal	8.2
HD 181440	B9 III	0.44 ± 0.05	0.975 ± 0.007	η Aql	Sec. Cal	7.5
HD 49968	K5 III	1.78 ± 0.02	0.939 ± 0.006	ζ Gem	Pri. Cal	4.1
HD 48450	K4 III	1.94 ± 0.02	0.949 ± 0.007	ζ Gem	Sec. Cal	9.5
HD 39587	G0 V	1.09 ± 0.04	0.963 ± 0.006	ζ Gem	Sec. Cal	16
HD 52711	G4 V	0.55 ± 0.04	0.962 ± 0.006	ζ Gem	Sec. Cal	8.8

Table 3.2: Relevant parameters of the calibrators. The angular separation listed is the angular distance from the calibrator to the Cepheid it is used to calibrate.

3.3   Analysis

3.3.1   Fringe Visibilities and Limb Darkening

PTI uses either a 10 or 20 ms sample rate. Each such sample provides a measure of the instantaneous fringe visibility and phase. While the phase value is converted to distance and fed back to the active delay line to provide active fringe tracking, the measured fringe visibility is averaged over the entire 130-second integration. The statistical uncertainty in each measurement is estimated by breaking the 130 second integration into five equal-time segments and measuring the standard deviation about the mean value.

The theoretical relation between source brightness distribution and fringe visibility is given by the van Cittert-Zernike theorem. For a uniform intensity disk model, the normalized fringe visibility (squared) can be related to the apparent angular diameter as

$$V^2 = \left(\frac{2\ J_1(\pi B\theta_{UD}/\lambda_0)}{\pi B\theta_{UD}/\lambda_0} \right)^2 \tag{3.1}$$

where J_1 is the first-order Bessel function, B is the projected aperture separation, θ_{UD} is the apparent angular diameter of the star in the uniform-disk model, and λ_0 is the center-band wavelength of the observation. It follows that the fringe visibility of a point source measured by an ideal interferometer should be unity. For a more realistic model that includes limb darkening one can derive a conversion factor between a uniform-disk diameter (θ_{UD}) and a limb-darkened disk diameter (θ_{LD}) given by Welch (1994)

$$\theta_{UD} = \theta_{LD} \sqrt{1 - \frac{A}{3} - \frac{B}{6}} \tag{3.2}$$

where A and B are quadratic limb darkening coefficients, determined by the spectral type of the source (Claret et al 1995). The limb darkening correction factors ($k = \theta_{UD}/\theta_{LD}$) used for the Cepheids are shown in Table 3.1 and for the calibrators in Table 3.2.

3.3.2 Visibility Calibration

The first step in calibrating visibilities measured by PTI is to correct for the effects of detector background and read-noise, the details of which are discussed in Colavita et al. (1999) and Colavita (1999b). However, the visibilities thus produced are not yet final: due to a variety of effects, including systematic instrumental effects, intensity mismatches, and atmospheric turbulence, the fringe visibility of a source measured by PTI is lower than that predicted by Eqn. 3.1 In practice the system response function (called the system visibility) is typically ~ 0.75 and furthermore is variable on 30 minute timescales. Hence the visibilities must be calibrated by observing sources of known diameter.

Determining the diameter of the calibration sources was a multi-step process in which we made use of both models and prior observations. For each Cepheid, we designated a single, bright K giant as a primary calibrator, which was always observed in close conjunction with the target Cepheid (HD 189695 for η Aql, and HD 49968 for ζ Gem). We used model diameter estimates for the primary calibrators from previously published results based on spectro-photometry and modeling (Cohen et al. 1999).

In order to verify that the primary calibrators were stable and had angular diameters consistent with the Cohen et al. (1999) results, we observed them together with a number of secondary calibrators. These secondary calibrators were typically less resolved than the primary calibrators and hence less sensitive to uncertainties in their expected angular diameter. However, they were fainter than the primary calibrators, and tended to be located further away on the sky. For the secondary calibrators an apparent diameter was estimated using three methods: (1) we used available archival photometry to fit a black-body model by adjusting the apparent angular diameter, bolometric flux and effective temperature of the star in question so as to fit the photometry. (2) We repeated the above fit while constraining the effective temperature to the value expected based on the published spectral type. (3) We estimated the angular diameter of the star based on expected physical size (derived from spectral type) and distance (determined by Hipparcos). We adopted the weighted (by the uncertainty in each determination) mean of the results from the above methods as the final model diameter for the secondary calibrators, and the uncertainty in the model diameter was taken to be the deviation about the mean.

In addition to the model-based diameter estimates derived above, we also used extensive interferometric visibility measurements for the primary and secondary calibrators; given that several of the calibrators were observed within a short enough period of time that the system visibility could be treated as constant, it was possible to find a set of assumed calibrator diameters that are maximally self-consistent, by comparing observed diameter ratios for which the system visibility drops out. To illustrate, let θ_i be an adjustable parameter, representing the diameter of star i. Let $\hat{\theta}_i$ and $\sigma_{\hat{\theta}_i}$ be the theoretical model diameter and uncertainty for star i derived above, and let $\tilde{R}_{ij}$ and $\sigma_{\tilde{R}_{ij}}$ be the interferometrically observed diameter ratio and uncertainty of stars i and j. For notational simplicity, define R_{ij} as the ratio of θ_i and θ_j. Define the quantity

$$\chi^2 = \sum_i \left[\frac{\hat{\theta}_i - \theta_i}{\sigma_{\hat{\theta}_i}} \right]^2 + \sum_i \sum_{j<i} \left[\frac{\tilde{R}_{ij} - R_{ij}}{\sigma_{\tilde{R}_{ij}}} \right]^2 \tag{3.3}$$

By adjusting the set of θ_i to minimize χ^2 we produce a set of consistent calibrator diameters, taking into account both input model knowledge and observations. The resulting diameter values are listed in Table 3.2. Uncertainties were estimated using the procedure outlined in Press et al. (1986) assuming normally distributed errors.

We verified that the primary calibrators were stable as follows: using the secondary calibrators to calibrate all observations of the primary calibrators we fit a constant-diameter, single-star, uniform-disk model to the primary calibrators. In all cases the scatter about the single-star model was similar to expected system performance (Boden et al. 1998): for HD 189695, 21 points were fit to an average deviation in V^2 of 0.035, and the goodness-of-fit parameter of the line fit, χ^2 per degree of freedom (χ^2_{dof}, not to be confused with Eq. 3 above), in the line fit was 0.46. For HD 49968, 82 points were fit, the average deviation was 0.038 and $\chi^2_{dof} = 0.76$.

While analyzing the data it was noticed that during observations with the N–W baseline of relatively low declination sources, such as η Aql and its calibrators, the stability of the interferometer system visibility was strongly dependent on the hour angle of the source: for observations of η Aql obtained at positive hour angles the scatter in the system visibility increased by a factor of 2–3, while the mean value trended down by 20%/hr. There are two potential explanations for this effect: (1) for these observations the optical delay lines are close to their maximum range, which can exacerbate internal system misalignments and lead to vignetting. (2) When observing low declination sources past transit, the siderostat orientation is such that surface damage near the edge of one of the siderostat mirrors causes vignetting. Thus it was decided to discard observations of η Aql taken at positive hour angles, corresponding to $\sim 20\%$ of the available data. We note that including the data does not significantly change the final results ($\sim 0.3\sigma$), it merely increases the scatter substantially (for the pulsation fit discussed below the goodness-of-fit parameter χ^2_{dof} increased from 1.06 to 4.5).

3.4 Results

3.4.1 Apparent Angular Diameters

Once the measured visibilities were calibrated we used all the available calibrated data from a given night to determine the apparent uniform-disk angular diameter of the target

Epoch JD-2400000.5	Angular Diameter θ_{UD} (mas)	No. Scans
52065.420	1.654 ± 0.011	9
52066.414	1.654 ± 0.017	9
52067.405	1.694 ± 0.040	8
52075.383	1.740 ± 0.027	12
52076.384	1.799 ± 0.014	9
52077.372	1.822 ± 0.021	13
52089.350	1.715 ± 0.019	11
52090.354	1.798 ± 0.020	9
52091.346	1.764 ± 0.022	7
52095.360	1.567 ± 0.049	1
52099.337	1.800 ± 0.025	2
52101.329	1.632 ± 0.037	5
52103.293	1.656 ± 0.040	7
52105.300	1.798 ± 0.024	6
52106.283	1.816 ± 0.016	19
52107.302	1.809 ± 0.027	11
52108.308	1.702 ± 0.032	7
52116.276	1.611 ± 0.023	7

Table 3.3: The measured uniform-disk diameters of η Aql. The uncertainties are the statistical uncertainty from the scatter during a night, and do not include systematic uncertainty in the calibrator diameters; this adds an additional uncertainty of 0.07 mas in the aggregate mean diameter.

Cepheid on that particular night by fitting to a model given by Eqn. 3.1 Results are given in Tables 3.3 and 3.4 and plotted in Figs. 3.1 and 3.2. Uncertainties were estimated based on the scatter about the best fit. It should be noted that although η Aql is known to have a companion (Bohm-Vitense & Proffitt 1985) it is sufficiently faint (average $\Delta m_H = 5.75$ mag) that it will have a negligible effect ($\Delta V^2 \sim 0.005$) on the fringe visibilities measured in the H band.

It is clear from Fig. 3.1 that the measured angular diameters are not constant with time. Fitting a constant-diameter model to the data produces a rather poor fit (see Table 3.5). However, we list the resulting mean angular diameters in order to facilitate comparison with previous interferometric results.

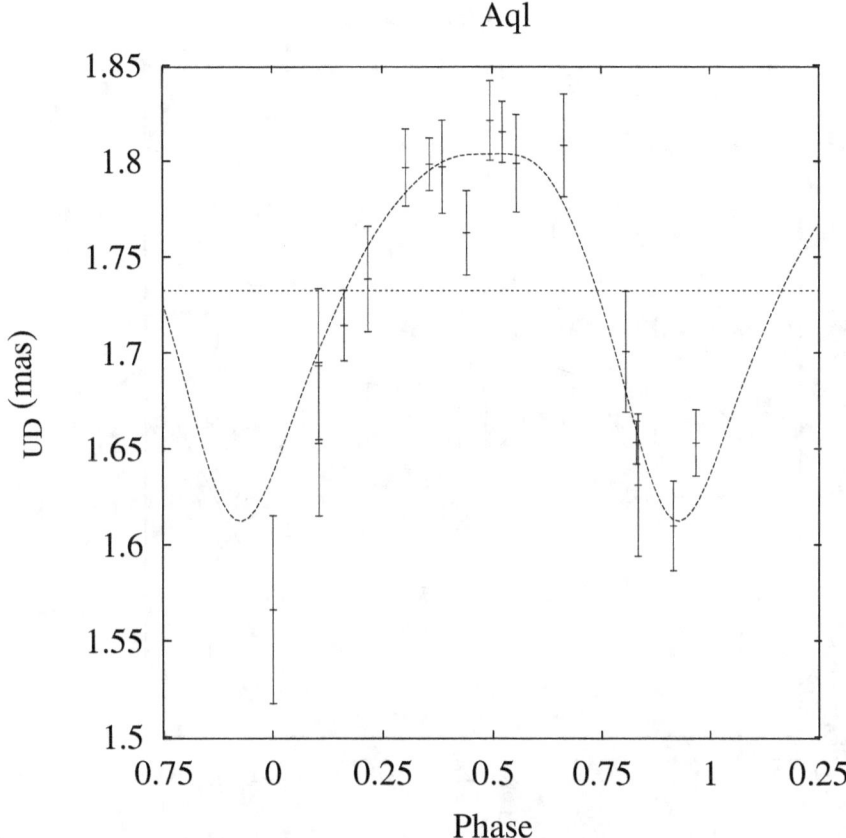

Figure 3.1: The angular diameter of η Aql as a function of pulsational phase, together with a model based on radial velocity data, but fitting for distance, mean radius and phase shift. Also shown is the result of fitting a line to all the data. The fits are extended past phase 0 for clarity.

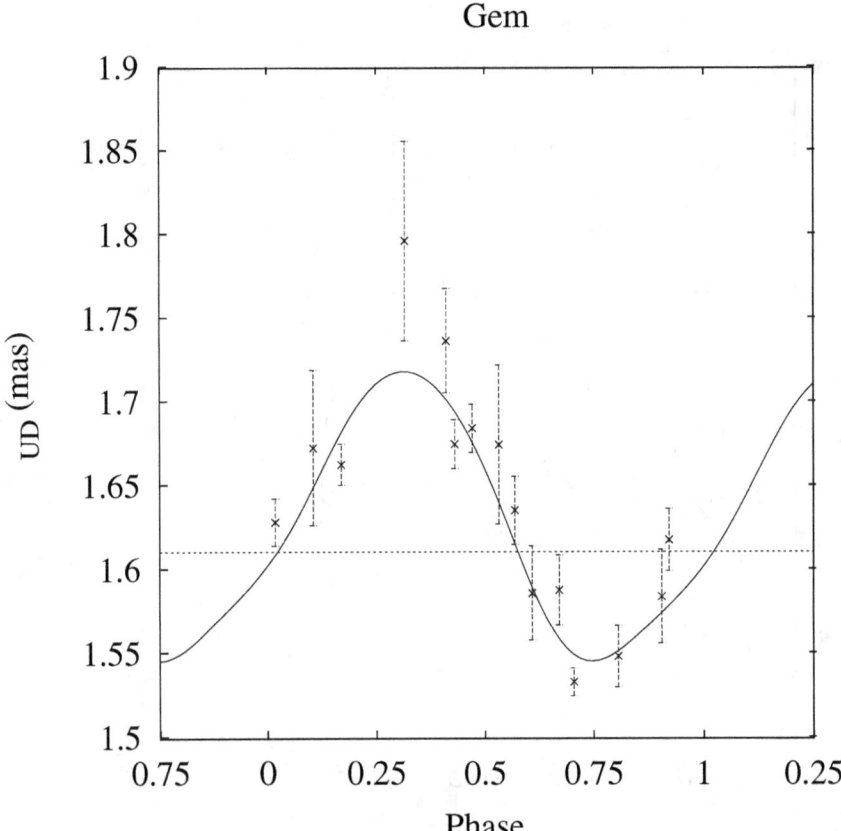

Figure 3.2: The angular diameter ζ Gem as a function of pulsational phase, together with a model based on radial velocity data, but fitting for distance, mean radius and phase shift. Also shown is the result of fitting a line to all the data. The fits are extended past phase 0 for clarity.

Epoch JD-2400000.5	Angular Diameter θ_{UD} (mas)	No. Scans
51605.226	1.676 ± 0.015	15
51606.241	1.675 ± 0.047	3
51614.192	1.797 ± 0.060	7
51615.180	1.737 ± 0.031	10
51617.167	1.587 ± 0.028	10
51618.143	1.534 ± 0.008	11
51619.168	1.549 ± 0.018	15
51620.169	1.585 ± 0.028	15
51622.198	1.673 ± 0.046	6
51643.161	1.663 ± 0.012	9
51981.182	1.685 ± 0.014	23
51982.164	1.636 ± 0.020	16
51983.201	1.589 ± 0.021	15
51894.387	1.619 ± 0.019	13
51895.369	1.629 ± 0.014	12

Table 3.4: The measured uniform-disk diameters of ζ Gem. The uncertainties are the statistical uncertainty from the scatter during a night, and do not include systematic uncertainty in the calibrator diameters; this adds an additional uncertainty of 0.024 mas in the aggregate mean diameter.

3.4.2 Distances & Radii

Determining the distance and radius of a Cepheid via the Baade-Wesselink method requires comparing the measured changes in angular diameter to the expansion of the Cepheid photosphere measured using radial velocity techniques. In order to determine the expansion of the Cepheid photospheres we fit a fifth-order Fourier series to previously published radial velocities. For η Aql we used data from Bersier (2002) as well as data published by Jacobsen & Wallerstein (1981) and Jacobsen & Wallerstein (1987), while for ζ Gem we used data from Bersier et al. (1994). Both sets of data were from measurements made at optical wavelengths. The measured radial velocities were converted to physical expansion rates using a projection factor (p-factor), which depends on the detailed atmospheric structure and limb darkening of the Cepheid as well as on the details of the equipment and software used in the measurement (Hindsley & Bell 1986, Albrow & Cottrell 1994) . It is important to note that the p-factor is not expected to stay constant during a pulsational cycle. The exact phase dependence of the p-factor is beyond the scope of this paper. However, for η Aql and ζ Gem, the net effect of a variable p-factor can be approximated by using a 6%

larger constant p-factor (Sabbey et al. 1995). Thus for both Cepheids we use an effective p-factor of 1.43 ± 0.06, constant for all pulsational phases.

We convert the radial velocity Fourier series into a physical size change by integrating and multiplying by limb-darkening and p-factors. Although the limb-darkening does vary with changing T_{eff} during a pulsational cycle, the effect is small: for ζ Gem k varies from 0.960 to 0.967, i.e. less than the quoted uncertainty. The size change can in turn be converted into an angular size model with three free parameters: the mean physical radius, the distance to the star, and a phase shift. The latter is to account for possible period changes, inaccuracies in period or epoch, or phase lags due to level effects (where the optical and infrared photospheres are at different atmospheric depths; see below). We adjust the model phase, radius and distance to fit the observed angular diameters. Results of the fits for η Aql and ζ Gem are given in Table 3.5.

There are several sources of uncertainty in the above fits: in addition to the purely statistical uncertainty there are systematic uncertainties of comparable magnitude. The three primary sources of systematic uncertainty are (1) uncertainty in the calibrator diameters, (2) uncertainty in the p-factor, and (3) uncertainty in the limb darkening coefficients. The magnitude of each effect was estimated separately by re-fitting the model while varying by $\pm 1\sigma$ each relevant parameter separately. The total systematic uncertainty was calculated as

$$\sigma_{sys}^2 = \sigma_{cal}^2 + \sigma_{p-fac}^2 + \sigma_{limbdark}^2. \tag{3.4}$$

In order to explore the possibility of wavelength-dependent effects on the measured radial velocity due to velocity gradients in the Cepheid atmospheres ("level effects"), we refit for the radius and distance of η Aql using a radial velocity curve based on radial velocity data obtained at wavelengths of 1.1 and 1.6μm by Sasselov & Lester (1990). Because of the limited number of observations available (e.g only 3 H-band measurements of η Aql) we used the shape of the radial velocity curve derived from the fit to the optical data (i.e. by using the same Fourier coefficients); the IR data was only used to determine an overall amplitude of the velocity curve. For the IR points we used an effective p-factor of 1.41 ± 0.03 as recommended by D. Sasselov (private communication) and based on an analysis by Sabbey et al. (1995), taking into account both the use of a constant p-factor and the use of parabolic line fitting. The resulting best-fit parameters are very similar to those

Cepheid	Fit Type	Parameter	Best-Fit Results
			Value $\pm\, \sigma_{Tot}$ ($\sigma_{Stat.}/\sigma_{Sys.}$)
η Aql	Pulsation Fit	Distance (D)	320 ± 32 (24/21) pc
	No. Pts. $= 18$	Radius (R)	61.8 ± 7.6 (4.5/6.1) $R_\odot$
	$\chi^2_{dof} = 1.06$	Phase (ϕ)	0.02 ± 0.011 ($0.011/5 \times 10^{-4}$) cycles
	Line Fit	θ_{UD}	$1.734 \pm 0.070(0.018/0.068)$ mas
	$\chi^2_{dof} = 13.4$		
ζ Gem	Pulsation Fit	Distance (D)	362 ± 38 (35/15) pc
	No. Pts. $= 15$	Radius (R)	66.7 ± 7.2 (6.3/3.4) $R_\odot$
	$\chi^2_{dof} = 1.82$	Phase (ϕ)	0.013 ± 0.016 ($0.016/3 \times 10^{-5}$) cycles
	Line Fit	θ_{UD}	$1.613 \pm 0.029(0.017/0.024)$ mas
	$\chi^2_{dof} = 14.6$		

Table 3.5: Best-fit Cepheid parameters and their uncertainties, as well as mean apparent uniform-disk angular diameter (θ_{UD}) determined from fitting a line to all of the data. The uncertainties of the best-fit parameters are broken down into statistical ($\sigma_{Stat.}$) and systematic ($\sigma_{Sys.}$) uncertainties. The goodness-of-fit parameter is a weighted χ^2 divided by the number of degrees of freedom (χ^2_{dof}) in the fit. The χ^2_{dof} of the fits are calculated from data that does not have the systematic (calibrator) uncertainty folded in since it applies equally to all points.

based on optical radial velocities (i.e., Table 3.5): D $= 333 \pm 30$ pc and R $= 64.2 \pm 6R_\odot$. A similar fit for ζ Gem gives D $= 359 \pm 37$ pc and R $= 62.2 \pm 5.7R_\odot$. Hence we conclude that the effects of wavelength dependence of the radial velocity are at present smaller than other sources of uncertainty.

The derived parameters (mean radius, distance and mean uniform-disk angular diameter) can be compared to previously published values, derived using a range of techniques (see Table 3.6), including parallax and a variety of surface brightness techniques. There are also several interferometric diameter measurements available in the literature, although to date no other interferometers have directly resolved Cepheid pulsations. Thus, directly measured angular diameters can only be compared in a phase-averaged sense.

3.4.3 Surface Brightness Relations

A wide variety of Cepheid surface brightness relations have been used by various authors (Barnes & Evans 1976, Laney & Stobie 1995, Fouque & Gieren 1997) to derive Cepheid

Cepheid	Reference	Radius R$_\odot$	Distance (pc)	Angular Diameter θ_{LD} (mas)
η Aql	this work	61.8 ± 7.6	320 ± 32	1.793 ± 0.070
	[Nordgren et al. 2000]			1.69 ± 0.04
	[Ripepi et al. 1997]	57 ± 3		
	[Perryman et al. 1997]		360^{+174}_{-89}	
	[Sasselov & Lester 1990]	62 ± 6		
	[Fernley, Skillen, & Jameson 1989]	53 ± 5	275 ± 28	
	[Moffett & Barnes 1987]	55 ± 4		
ζ Gem	this work	66.7 ± 7.2	362 ± 38	1.675 ± 0.029
	[Lane et al. 2000]	62 ± 11	336 ± 44	1.62 ± 0.3
	[Kervella et al. 2001]			$1.69^{+0.14}_{-0.16}$
	[Nordgren et al. 2000]			1.55 ± 0.09
	[Perryman et al. 1997]		358^{+147}_{-81}	
	[Ripepi et al. 1997]	86 ± 4		
	[Bersier et al. 1997]	89.5 ± 13	498 ± 84	
	[Krockenberger, Sasselov, & Noyes 1997]	$69.1^{+5.5}_{-4.8}$		
	[Sabbey et al. 1995]	64.4 ± 3.6		
	[Moffett & Barnes 1987]	65 ± 12		

Table 3.6: A comparison between the various available radius, distance and angular size determinations. The Nordgren et al. (2000) results are based on R band (740 nm) observations, while the Kerevella et al. (2001) result is in the K band (2.2 μm).

Cepheid	A_V	A_R	A_K
η Aql	0.515	0.377	0.055
ζ Gem	0.062	0.046	0.007

Table 3.7: Reddening values used in deriving surface brightness parameters for η Aql and ζ Gem, based on values of $E(B-V)$ from Fernie (1990).

distance scales. We define as surface brightness the quantity

$$F_i = 4.2207 - 0.1 m_i - 0.5 \log(\theta_{LD}) \qquad (3.5)$$

where F_i is the surface brightness in magnitudes in passband i, m_i is the apparent magnitude in that band, and θ_{LD} is the apparent angular diameter of the star. With the above relation and a good estimate of F_i one can determine the angular diameter based on photometry alone. Conversely, given measured angular diameters and multi-band photometry it is possible to calibrate F_i by finding a simple (e.g. linear) relation between F_i and a variety of color indices (e.g. $V - K$). We define the following relations

$$F_{V,1} = a + b(V - K) \qquad (3.6)$$

$$F_{V,2} = a + c(V - R) \qquad (3.7)$$

Note that consistency requires a common zero-point (cf. an A0V star where $(V - R) = (V - K) = 0$).

We used previously published VRK photometry of η Aql (Barnes et al. 1997) to derive its apparent magnitude in the above bands as a function of phase by fitting a low-order Fourier series to the published photometry, after first correcting for the effects of reddening following the procedure outlined in Evans & Jiang (1993). The individual values of $E(B-V)$ were taken from Fernie (1990), and the reddening corrections applied are listed in Table 3.7. For each diameter measurement we then used the Fourier series to derive m_V and $V - K$ at the epoch of observation, and using Eqn. 3.5 we derived the corresponding surface brightness. Results are shown in Figure 3.4 and listed in Table 3.8. We also performed this type of fit using ζ Gem data. In this case we used photometry from Wisniewski & Johnson (1968) and Moffett & Barnes (1984).

Source	a	b	c
η Aql, this work	3.941 ± 0.005	-0.125 ± 0.004	-0.375 ± 0.002
ζ Gem, this work	3.946 ± 0.011	-0.130 ± 0.002	-0.378 ± 0.003
[Fouque & Gieren 1997]	3.947 ± 0.003	-0.131 ± 0.003	-0.380 ± 0.003
[Nordgren et al. 2000]	3.941 ± 0.004	-0.125 ± 0.003	-0.368 ± 0.007

Table 3.8: A comparison between the various surface brightness relations (see text for definitions).

In Table 3.8 we compare the derived surface brightness relations to similar relations from work based on non-variable supergiants (Fouque & Gieren) and other Cepheid observations (Nordegren et al. 2001). The F_V vs. $V - R$ fits can also be compared with the Gieren (1988) result that the slope of the $V - R$ surface brightness relation (c) is weakly dependent on pulsational period (P) according to

$$c = -0.359 - 0.020 \log P \qquad (3.8)$$

which for η Aql predicts $c = -0.376$ and for ζ Gem $c = -0.379$. These comparisons reveal generally good agreement between the various relations in Table 3.8.

3.4.4 Period-Radius Relations

The relation between pulsational period and Cepheid radius has received considerable attention in the literature, primarily because early results based on different techniques were discrepant (Fernie 1984, Moffett & Barner 1987). Period-radius relations are also useful in that they can indicate pulsation mode. This is important for calibrating period-luminosity relations since different modes will yield different relations (Feast & Catchpole 1997, Nordgren et al. 2001).

In Fig. 3.5 we compare our measured Cepheid diameters to the values predicted from a range of techniques: Bono, Capute & Marconi (1998) calculate a period-radius relation from full-amplitude, nonlinear, convective models for a range of metallicities and stellar masses. Gieren, Moffett & Barner (1999) use the surface brightness technique based on V and $V - R$ photometry and the Fouque & Gieren (1997) result to derive radii for 116 Cepheids in the Galaxy and the Magellanic Clouds. They find an intrinsic width in their relation of ± 0.03 in log R. Laney & Stobie (1995) also use the surface brightness technique for estimating

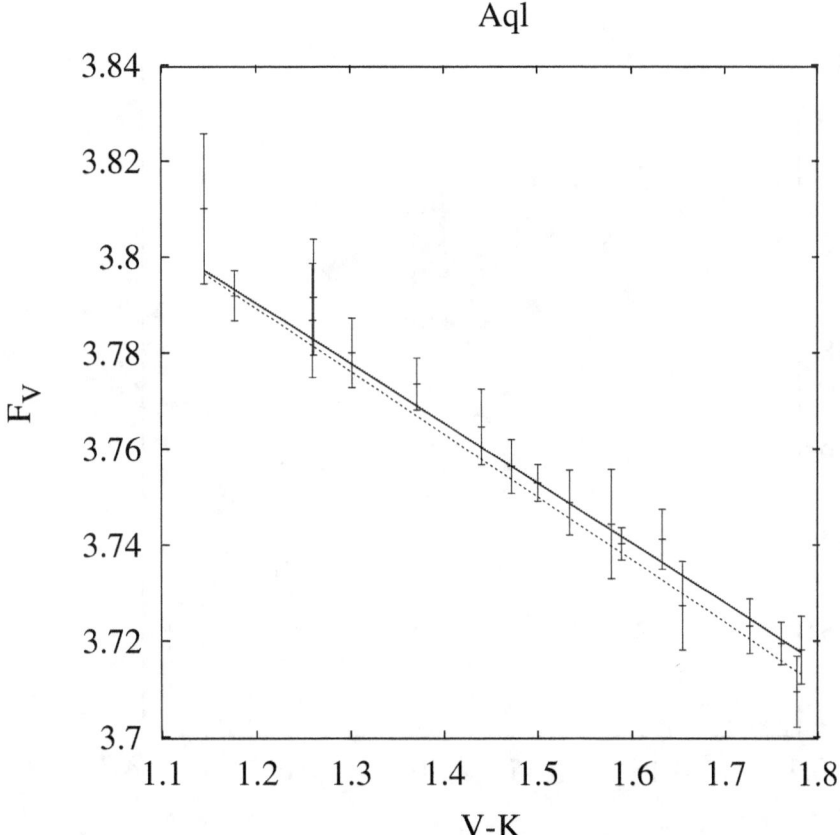

Figure 3.3: Dereddened F_V vs. $V - K$ for η Aql. The solid line is the weighted linear least-squares fit to the data. The dashed line represents the relation from Foque & Gieren (1997), and the dotted line represents the Nordgren et al. (2001) result.

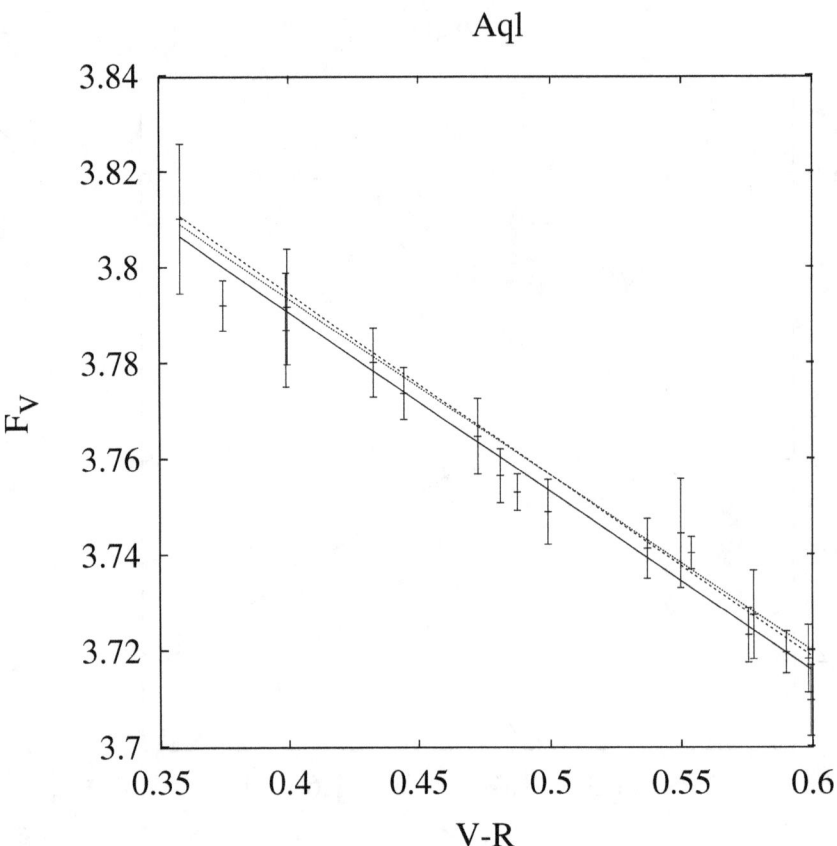

Figure 3.4: Dereddened F_V vs. $V - R$ for η Aql. The solid line is the weighted linear least-squares fit to the data. The dashed line represents the relation from Foque & Gieren (1997), and the dotted line represents the Nordgren et al. (2001) result.

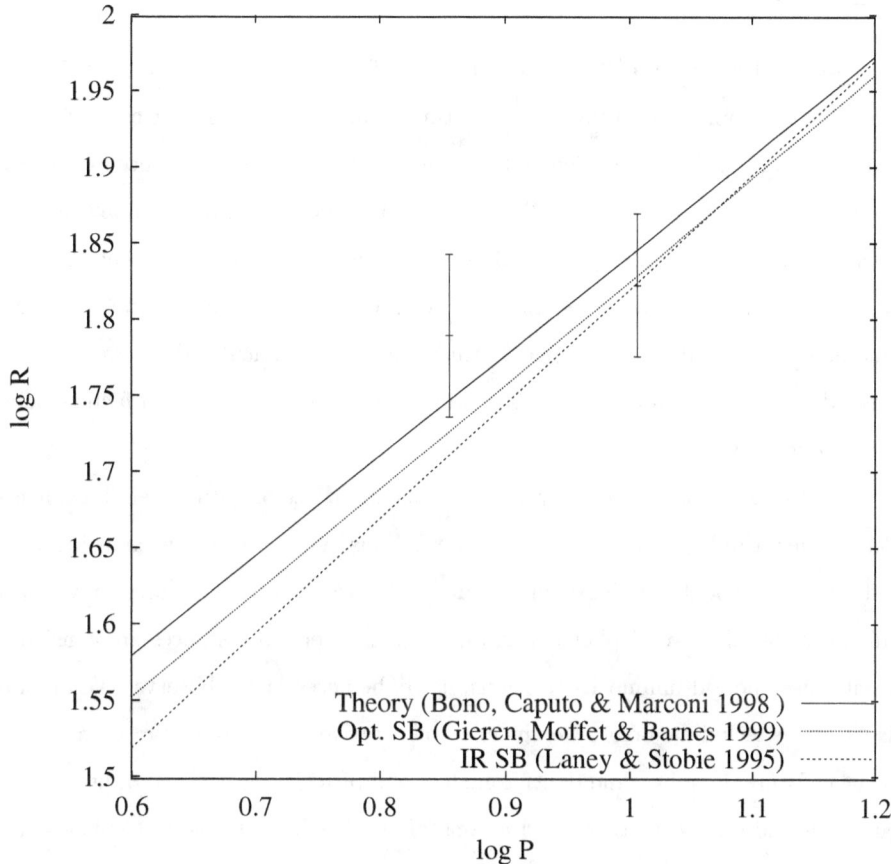

Figure 3.5: Period-radius diagram for the two Cepheids η Aql and ζ Gem, together with three relations available in the literature: a theoretical relation derived by Bono, Caputa & Marconi (1998), an optical surface brightness relation from Gieren, Moffett & Barnes (1999) and an IR surface brightness relation from Laney & Stobie (1995).

Cepheid diameters. However, they find that infrared photometry $(K, J - K)$ is less sensitive to the effects of gravity and microturbulence (and presumably also reddening), and hence yields more accurate results. For shorter periods (≤ 11.8 days) their results indicate smaller diameters as compared to other relations.

Given the limited sample of only two radius measurements we can draw only preliminary conclusions: (1) the general agreement between our observations and the relations is good, and (2) the data seem to prefer a shallower slope than the Laney & Stobie (1995) relation. This latter observation will have to be confirmed with observations of shorter-period Cepheids.

3.5   Summary

We have measured the changes in angular diameter of two Cepheids, η Aql and ζ Gem, using PTI. When combined with previously published radial velocity data we can derive the distance and mean diameter to the Cepheids. We find η Aql to be at a distance of 320 ± 32 pc with a mean radius of $61.8 \pm 7.6 R_\odot$. We find ζ Gem to be at a distance of 362 ± 38 pc, with a mean radius of $66.7 \pm 7.2 R_\odot$, in good agreement with previous work. The precision achieved is $\sim 10\%$ in the parameters; further improvement is at present limited by our understanding of the details of the Cepheid atmospheres. In particular the details of limb darkening and projection factors need to be understood, with the projection factors being the largest source of systematic uncertainty.

We note that these results do not rely on photometric surface brightness relations, hence results derived here can be used to calibrate such relations. We performed such calibrations and found good agreement with previous results. We also note that at present we have derived distances to only two Cepheids, and although the derived distances are consistent with currently used period-luminosity relations, it will be necessary to observe several more Cepheids with this technique before worthwhile quantitative comparisons can be made.

In the near future, long-baseline interferometers will provide a great deal of useful data in this area: in addition to further observations of the brightest Galactic Cepheids, the very long baselines currently being commissioned at the Navy Prototype Optical Interferometer (Armstrong et al. 2001b) and the Center for High Angular Resolution Astronomy array (ten Brummelaar et al. 2001) will allow direct measurements of the limb darkening effects through observations of fringe visibilities past the first visibility null. Given the close relation between limb darkening and projection factors we expect that improvements in understanding one will improve our understanding of the other. It is also clear that additional photometry and radial velocity measurements would be very useful. In particular ζ Gem suffers from a lack of good infrared photometry, while concerns about level effects make infrared radial velocity measurements like those of Sasselov & Lester very desirable.

Chapter 4

Binary Star Observations

We report on the determination of the visual orbit of the double-lined spectroscopic binary system BY Draconis with data obtained by the Palomar Testbed Interferometer in 1999–2002. BY Dra is a nearly equal-mass double-lined binary system whose spectroscopic orbit is well known. We have estimated the visual orbit of BY Dra from our interferometric visibility data fit both separately and in conjunction with archival radial velocity data. Our BY Dra orbit is in good agreement with the spectroscopic results.

4.1   Introduction

BY Draconis (HDE 234677, Gl 719) is a well-studied, nearby ($\sim$ 15 pc), multiple stellar system containing at least three objects. The A and B components form a short-period (6 d) late-type (K6 Ve – K7 Vvar) binary system, whose spectroscopic orbit is well known (Bopp & Evans 1973, hereafter BE73; Vogt & Fekel (1979), hereafter VF79; and Lucke & Mayor (1980), hereafter LM80). BY Dra is the prototype of a class of late-type flare stars characterized by photometric variability due to star spots, rapid rotation, and Ca II H and K emission lines. Like the BY Dra system itself, a large fraction (> 85%, Bopp & Fekel 1977, Bopp et al. 1980) of BY Dra stars are known to be in short-period binary orbits. The rapid rotation of BY Dra A (period 3.83 d) that gives rise to the spotting and photometric variability is consistent with pseudosynchronus rotation with the A – B orbital motion (Hut 1981, Hall 1986), but pseudosynchronization is disputed by Glebocki & Stawikowski (1995, 1997) who assert asynchronous rotation and roughly 30° misalignment of the orbital/rotational angular momentum vectors.

---

[a]The material in this chaper was previously published as Boden & Lane, ApJL, 547, 1071B ,2001.

Despite the fact that the BY Dra A and B components are nearly equal mass, the system exhibits a significant brightness asymmetry in spectroscopic studies (VF79, LM80). VF79 attributes this to the hypothesis that A and B components of BY Dra are pre-main sequence objects (VF79, Bopp et al., 1980), and are still in the contraction phase. VF79 argues for physical sizes of the A component in the range of 0.9 – 1.4 $R_\odot$, based primarily on rotation period and $v \sin i$ considerations. LM80 concur with the A component physical size argument from their $v \sin i$ measurements, estimating a 1.2 $R_\odot$ size for a $\sin i \approx$ 0.5 (presuming rotation/orbit spin alignment with pseudosynchronization, and our orbital inclination from Table 4.2). However, they continue by pointing out that if the A component macroturbulance were significantly larger than solar, then the $v \sin i$ measurements and the component diameters they are based on are biased high. If the pre-main sequence interpretation is correct, the BY Dra components are additionally interesting as an examples of the transition region between pre and zero-age main sequence states.

BY Dra was detected as a hierachical triple system through common proper motion measurements of a BY Dra C component by Zuckerman et al. (1997); they find the C component separated by 17" from the A – B pair. Zuckerman's photometry on the C component is consistent with an M5 main-sequence interpretation (V - $K \approx$ 6.2); assuming all three stars are coeval this clearly poses problems for the VF79 pre-main sequence hypothesis for the A and B components. At a projected physical separation of approximately 260 AU from the A – B binary, the putative low-mass C component would have negligible dynamical influence on the A – B binary motion. Further, the Hipparcos catalog (ESA 1997) implies the presence of at least one additional component as it lists BY Dra as having a circular photocentric orbital solution with 114 d period. This 114 d period is previously unreported in spectroscopic studies, and if correct it is difficult to understand why this motion was not previously detected.

Herein we report on a preliminary determination of the BY Dra A – B system visual orbit from near-infrared, long-baseline interferometric measurements taken with the Palomar Testbed Interferometer (PTI). PTI is a 110-m H (1.6μm) and K-band (2.2 μm) interferometer located at Palomar Observatory, and described in detail elsewhere (Colavita et al. 1999). PTI has a minimum fringe spacing of roughly 4 milliarcseconds (10^{-3} arcseconds, mas) in K-band at the sky position of BY Dra, allowing resolution of the A – B binary system.

4.2 Observations

The interferometric observable used for these measurements is the fringe contrast or *visibility* (squared, V^2) of an observed brightness distribution on the sky. As derived in Chapter 1, the fringe visibility of a binary star is given by

$$|\hat{V}|^2 = \frac{1}{(1+R)^2}\left(V_1^2 + V_2^2 R^2 + 2V_1 V_2 R \cos(\frac{2\pi}{\lambda}\Delta\vec{s}\cdot\vec{B})\right) \tag{4.1}$$

where V_i is the uniform-disk visibility of the ith component from Eqn. 1.36, the intensity ratio $R = I_1/I_0$ and the binary separation vector is $\Delta\vec{s}$.

BY Dra was observed in conjunction with objects in our calibrator list (Table 4.1) by PTI in K-band ($\lambda \sim 2.2\mu$m) on 36 nights between June 23, 1999 and October 24, 2002, covering roughly 200 periods of the system. Additionally, BY Dra was observed by PTI in H-band ($\lambda \sim 1.6\mu$m) on 5 nights in 1999 through 2001. BY Dra, along with calibration objects, was observed multiple times during each of these nights, and each observation, or scan, was approximately 130 sec long. For each scan we computed a mean V^2 value from the scan data, and the error in the V^2 estimate from the rms internal scatter. BY Dra was always observed in combination with one or more calibration sources within $\sim 10°$ on the sky. Table 4.1 lists the relevant physical parameters for the calibration objects. We have calibrated the V^2 data by methods discussed in Boden et al. (1998). Our observations of BY Dra result in 170 calibrated visibility measurements (136 in K-band, 34 in H-band). One notable aspect of our BY Dra observations is that its high declination (51°) relative to our Palomar site (33° latitude) puts it at the extreme Northern edge of the delay line range on our N-S baseline, implying extremely limited $u - v$ coverage on BY Dra. To our PTI visibilities we have added 44 double-lined radial velocity measurments: 14 from BE73, seven from VF79, and 23 CORAVEL measurements from LM80.

4.3 Orbit Determination

The estimation of the BY Dra visual orbit is made by fitting a Keplerian orbit model directly to the calibrated (narrow-band and synthetic wide-band) V^2 and RV data on BY Dra; because of the limited $u - v$ coverage in our data derivation of intermediate separation vector models is impossible. The fit is non-linear in the Keplerian orbital elements, and is

Object Name	Spectral Type	Star Magnitude	BY Dra Separation	Adopted Model Diameter (mas)
HD 177196	A7 V	5.0 V/4.5 K	6.6°	0.70 ± 0.06
HD 185395	F4 V	4.5 V/3.5 K	9.9°	0.84 ± 0.04

Table 4.1: PTI BY Dra calibration objects considered in our analysis. The relevant parameters for our two calibration objects are summarized. The apparent diameter values are determined from effective temperature and bolometric flux estimates based on archival broad-band photometry, and visibility measurements with PTI.

therefore performed by non-linear least-squares methods with a parallel exhaustive search strategy to determine the global minimum in the chi-squared manifold.

Figure 4.1 depicts the relative visual orbit of the BY Dra system, with the primary component rendered at the origin, and the secondary component rendered at periastron. We have indicated the phase coverage of our V^2 data on the relative orbit with heavy lines; our data samples most phases of the orbit well, leading to a reliable orbit determination. The apparent inclination is very near the estimate given by VF79 based on the primary rotation period, assumed size, and assumption of parallel orbital/rotational angular momentum alignment. The orbit is seen approximately 30° from a face-on perspective, which makes physical parameter determination difficult (Sec. 4.5).

Figure 4.2 illustrates comparisons between our PTI V^2 and archival RV data and our orbit model. In Figure 4.2a six consecutive nights of K-band visibility data and visibility predictions from the best-fit model are shown, inset with fit residuals along the bottom. That there are only 1–3 data points in each of the nights follows from the brief time each night that BY Dra is simultaneously within both delay and zenith angle limits. Figure 4.2b gives the phased archival RV data and model predictions, inset with a histogram of RV fit residuals. The fit quality is consistent with previous PTI orbit analyses.

Spectroscopic orbit parameters (from VF79 and LM80) and our visual and spectroscopic orbit parameters of the BY Dra system are summarized in Table 4.2. We give the results of separate fits to only our V^2 data (our "V^2-only Fit" solution), and a simultaneous fit to our V^2 data and the archival double-lined radial velocities – both with component diameters constrained as noted above. All uncertainties in parameters are quoted at the one sigma level. We see good statistical agreement between all the derived orbital parameters with the exception of the LM80 period estimate.

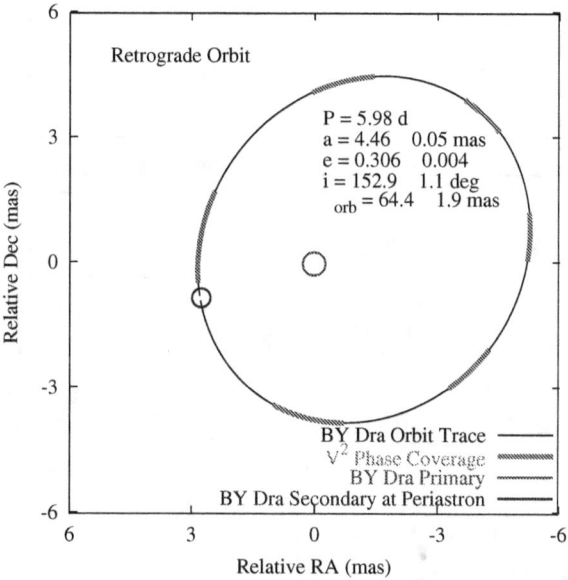

Figure 4.1: The relative visual orbit model of BY Dra is shown, with the primary and secondary objects rendered at T_0 (periastron). The heavy lines along the relative orbit indicate areas where we have orbital phase coverage in our PTI data (they are not separation vector estimates); our data samples most phases of the orbit well, leading to a reliable orbit determination. Component diameter values are estimated (see discussion in Sec. 4.5), and are rendered to scale.

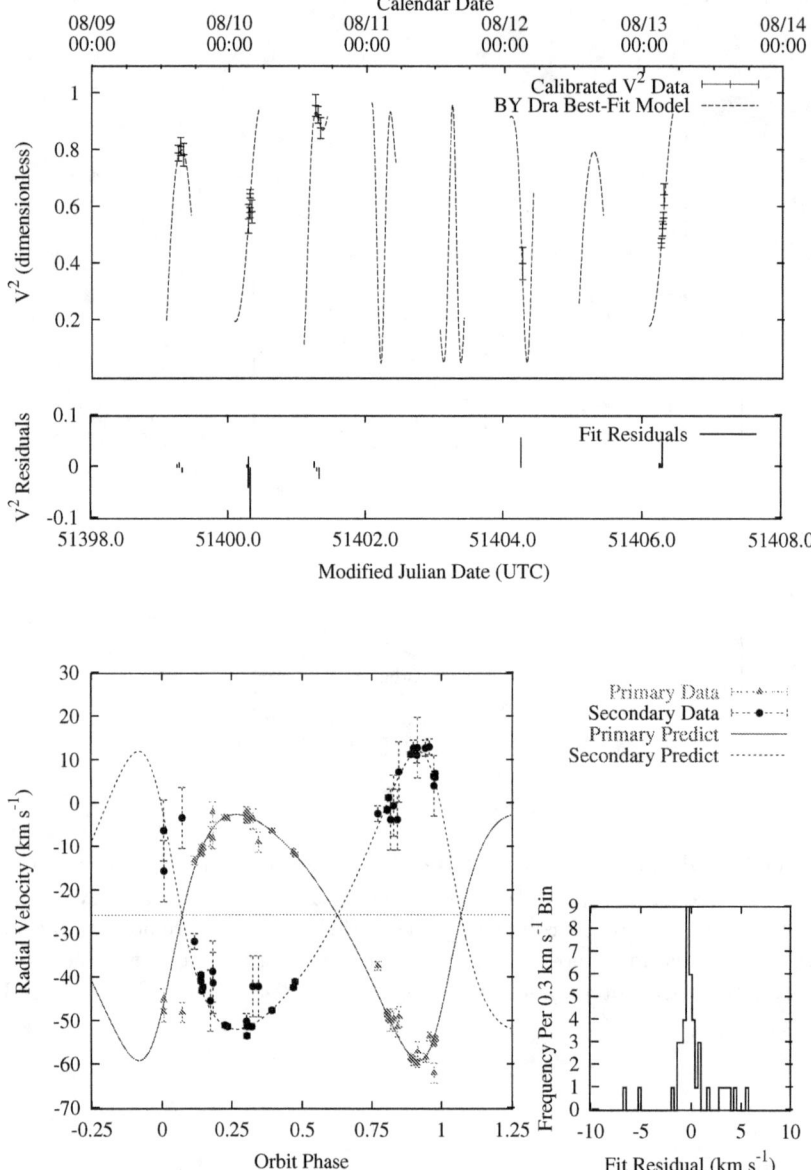

Figure 4.2: a: Five nights of V^2 data on BY Dra, and best-fit model predictions. b: Phased archival RV data (BE73, VF79, LM80) and RV predictions from our best-fit orbital model. Inset is a histogram of RV residuals to the fit model.

Orbital Parameter	VF79	LM80	PTI 99	
			V^2-only Fit	Full Fit
Period (d)	5.9750998	5.975112	*5.975078*	5.975107
	$\pm\, 8.7 \times 10^{-5}$	$\pm\, 1.2 \times 10^{-5}$		$\pm\, 1.1 \times 10^{-5}$
T_0 (MJD)	41146.59	43794.193	51376.1356	51376.1309
	$\pm\, 0.07$	$\pm\, 0.023$	$\pm\, 0.007$	$\pm\, 0.005$
e	0.36	0.3066	0.307	0.306
	$\pm\, 0.03$	$\pm\, 0.0063$	$\pm\, 0.003$	$\pm\, 0.002$
K_1 (km s^{-1})	28.2 ± 1.0	28.55 ± 0.25		28.38 ± 0.22
K_2 (km s^{-1})	28.8 ± 1.8	32.04 ± 0.35		32.07 ± 0.12
γ (km s^{-1})	-24.47 ± 0.65	-25.35 ± 0.14		-25.431 ± 0.128
ω_1 (deg)	220 ± 5	229.3 ± 1.3	228.8 ± 1.2	230.0 ± 1.0
Ω_1 (deg)			118.9 ± 2.3	117.8 ± 1.0
i (deg)			152.9 ± 1.5	152.9 ± 1.1
a (mas)			4.452 ± 0.033	4.460 ± 0.050
ΔK_{CIT}			0.558 ± 0.025	0.512 ± 0.061
ΔH_{CIT}			0.336 ± 0.30	0.336 ± 0.24
ΔV		1.15 ± 0.1		
χ^2/DOF			2.1	2.6 (2.2 V^2/3.6 RV)
$\overline{|R_{V^2}|}$			0.044	0.044
$\overline{|R_{RV}|}$ (km s^{-1})		0.55		2.3 (0.49 COR)

Table 4.2: Summarized here are the apparent orbital parameters for the BY Dra system as determined by VF79, LM80, and PTI. We give two separate fits to our data, with and without including archival double-lined radial velocities in the fit. Quantities given in italics are constrained to the listed values in our model fits. We have quoted the longitude of the ascending node parameter (Ω) as the angle between local East and the orbital line of nodes measured positive in the direction of local North. Due to the degeneracy in our V^2 observable there is a 180° ambiguity in Ω; by convention we quote it in the interval of [0:180). We quote mean absolute V^2 and RV residuals in the fits, $\overline{|R_{V^2}|}$ and $\overline{|R_{RV}|}$, respectively.

4.4   Comparisons with Hipparcos Model

The Hipparcos catalog lists a circular photocentric orbital solution for BY Dra with a 114 day period (ESA 1997), presumably in addition to the well-established 6 d period A–B motion. As noted above it is difficult to reconcile this hypothesis with the quality of the existing short-period spectroscopic orbit solutions from VF79 and LM80. However if the A–B system indeed did have a companion with this period, unlike BY Dra C it would lie within PTI's 1" primary beam, and if sufficiently luminous it would bias the visibility measurements used in our BY Dra A–B visual orbit model. We see no indications of this in our orbital solutions; the quality of our BY Dra visual orbit solution is consistent with our results on other systems. But it remains possible we have misinterpreted our V^2 data in the binary star model fit, and we are motivated to consider the 114-d periodicity hypothesis in the archival RV data.

We note that the Hipparcos model is a photocentric orbit, and therefore calls for the A − B system to exhibit a reflex motion with radius ≥ 0.05 AU at the putative distance of BY Dra. The 114-d orbit hypothesis is at high inclination (113°), and therefore would produce a radial velocity semi-amplitude for the A − B system barycenter ≥ 3.95 km s^{-1}. This value is large compared to fit residuals observed by VF79 and LM80 in their spectroscopic orbital solutions (and by ourselves in the joint fit with our visibilities; Table 4.2), suggesting the 114-d motion hypothesis is unlikely.

To quantify this issue we have considered Lomb-Scargle periodogram analyses of the LM80 BY Dra radial velocity data. We have chosen to use the LM80 data because it is the more precise sample, yielding an rms residual of roughly 0.5 km s^{-1} in our A–B orbit analysis. Figure 4.3 gives periodograms of the LM80 primary and secondary radial velocity data. First, Figure 4.3a gives a periodogram of the primary and secondary RV data, with probability of false alarm levels (P_{fa}) as noted. As expected, both component lines exhibit a significant periodicity at the observed A − B orbit frequency of approximately (6 d)$^{-1}$ (indicated by vertical line). In the same plot we sample the frequency of the 114-d orbit hypothesis, and no comparable periodicity is evident on the scale of the A − B motion.

Presuming the 114-d motion hypothesis might be superimposed on the 6-d A − B orbit, in Figure 4.3b we give a periodogram of the LM80 primary and secondary RV fit residuals to the 6-d hypothesis fit. In Figure 4.3b we have adjusted to range of the periodogram to

Physical Parameter	Primary Component	Secondary Component
a (10^{-2} AU)	3.25 ± 0.12	3.68 ± 0.14
Mass ($M_\odot$)	0.659 ± 0.053	0.583 ± 0.047
System Distance (pc)	15.5 ± 0.4	
π_{orb} (mas)	64.4 ± 1.9	
Model Diameter (mas)	*0.60 ($\pm$ 0.06)*	*0.50 ($\pm$ 0.05)*
M_K (mag)	4.44 ± 0.07	4.95 ± 0.07
M_H (mag)	4.64 ± 0.13	4.98 ± 0.16
M_V (mag)	7.44 ± 0.07	8.59 ± 0.08
V-K (mag)	2.997 ± 0.033	3.635 ± 0.048

Table 4.3: Summarized here are the physical parameters for the BY Dra A – B system as derived primarily from the Full-Fit solution orbital parameters in Table 4.2. Quantities listed in italics (i.e., the component diameters, see text discussion) are constrained to the listed values in our model fits.

finely sample the frequency range around the $(114 \text{ d})^{-1}$ hypothesis (indicated by vertical line). No significant periodicity is noted at this or any other frequency. The LM80 RV dataset spans roughly 400 days, so a 4 km s^{-1} amplitude periodicity in this dataset should have been evident in our analysis. Given these considerations, it seems unlikely that our V^2 measurements and A – B orbit model are affected by the presence of a third luminous body.

4.5 Discussion

The combination of the double-lined spectroscopic orbit and relative visual orbit allow us to estimate the BY Dra A – B component masses and system distance. However, because of the nearly face-on geometry of the A – B orbit, the accuracy of our inclination estimate, and consequently the component mass estimates and system distance estimate, is only $\sim 8\%$. The low-inclination geometry is particularly difficult for astrometric studies because the astrometric observable (V^2 in this case) becomes highly insensitive to small changes in the inclination Euler angle. Table 4.3 lists the physical parameters we derive from our Full-Fit orbit solution. The orbital parallax is in 1-σ agreement with the Hipparcos triginometric determination of 60.9 ± 0.75 mas, but we note that the Hipparcos solution was derived jointly with the 3 mas, 114-d orbital hypothesis that we believe to be suspect (§4.4).

Table 4.3 also gives component absolute magnitudes and V - K color indices derived from archival broad-band photometry, our H and K component relative magnitudes, the

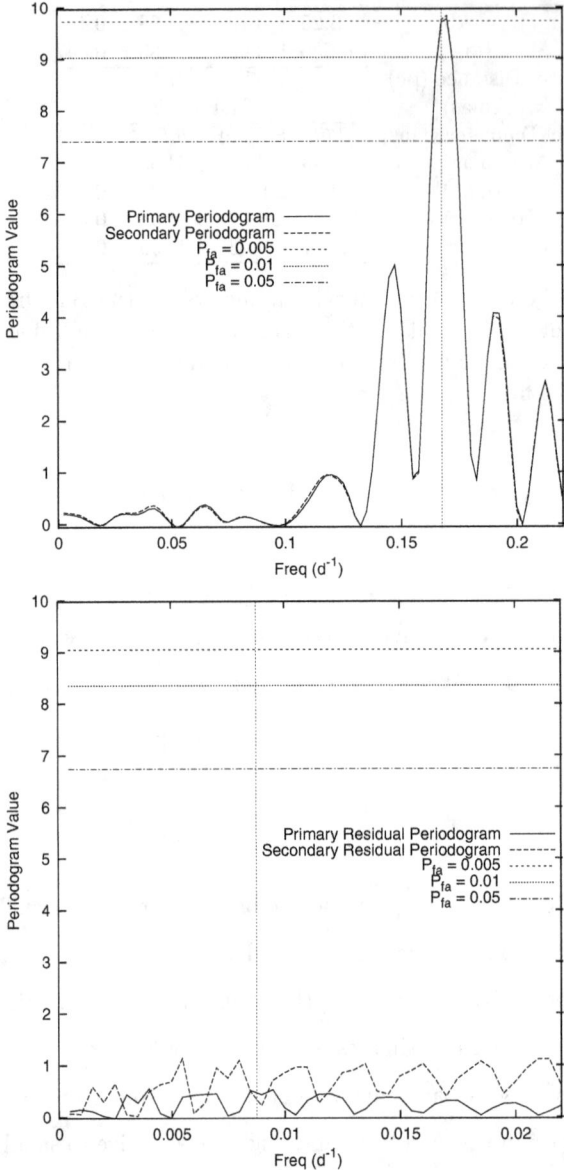

Figure 4.3: BY Dra Radial Velocity Data. To assess the 114-d orbit hypothesis for an additional companion to the BY Dra A – B system we have performed Lomb – Scargle periodogram analysis on double-lined BY Dra data from LM80. Top: standard periodograms on primary and secondary RV data from the LM80 dataset. Statistically significant response is seen at the well-known 6 d A – B period (vertical line). Bottom: Low-frequency periodogram of the residuals of the LM80 data to the best-fit A – B orbit model. No significant response is evident at the putative 114 d period (vertical line).

LM80 V relative magnitude, and our system distance estimate. The component color indices are unaffected by errors in the system distance, and seem to be in good agreement with the color indices expected from stars in this mass range and the classical spectral typing of the BY Dra system.

To assess the VF79 component size/pre-main sequence hypothesis, the most interesting measurement of the BY Dra A – B system would be unequivocal measurements of the component diameters. Unfortunately our V^2 data are as yet insufficient to determine these diameters independently. Canonical sizes of 1.0 $R_\odot$ and 0.8 $R_\odot$ (indicated by model effective temperatures and our IR flux ratios) yield angular diameters of approximately 0.6 and 0.5 mas for the A and B components respectively. At these sizes neither the H nor K-band fringe spacings of PTI sufficiently resolve the components to independently determine component sizes. Consequently we have constrained our orbital solutions to these 0.6 and 0.5 mas model values. Our data does in fact prefer the slightly smaller primary component diameter to the larger 1.2 $R_\odot$ size implied by VF79 and LM80 $v \sin i$ and rotational period measurements. However either primary model diameter is possible with the expected systematic V^2 calibration errors. Additional data we will collect in the coming year may well place interesting upper limits on the component sizes, but unambiguous resolution of the BY Dra A and B components will have to wait for a longer baseline infrared interferometer; most likely the CHARA array currently under construction on Mt. Wilson [a]

---

[a]http://www.chara.gsu.edu/CHARAArray/chara_array.html

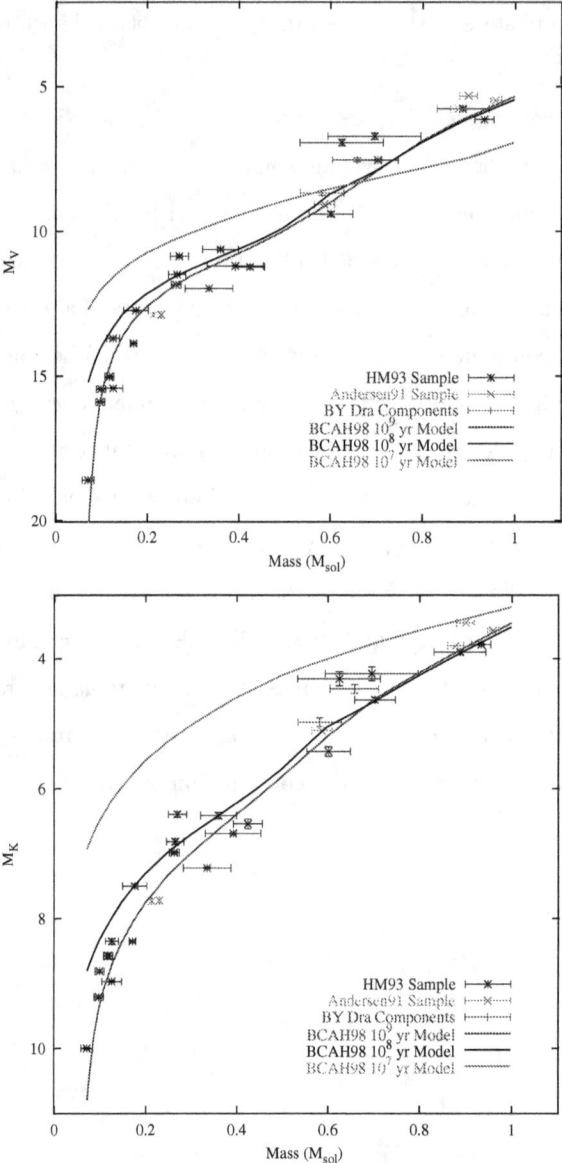

Figure 4.4: Here we give the positions of the A and B BY Dra components in observable Mass/M_V (upper pannel) and Mass/M_K (lower pannel) spaces (an emulation of Figure 3 from BCAH98), superimposing low mass objects with masses known to better than 20 %, from Henry & McCarthy (1993) and Andersen (1990) and BCAH98 solar-metalicity model tracks at ages of 10^9, 10^8, and 10^7 yrs.

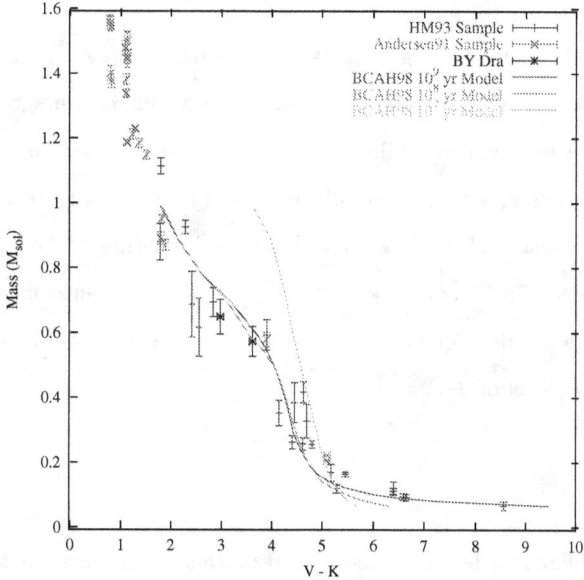

Figure 4.5: Here we give the positions of the A and B BY Dra components in Mass vs. V–K color space, superimposing low mass objects from Henry & McCarthy (1993) and Andersen (1990) and BCAH98 solar-metalicity model tracks at ages of 10^9, 10^8, and 10^7 yrs.

Chapter 5

Phase Referencing and Astrometry

We discuss implementation and testing of phase referencing at the Palomar Testbed Interferometer (PTI). A new instrument configuration provides a coherent integration of 10 or 20 milliseconds on a bright star while stabilizing the fringe phase of a nearby (20 arcseconds) and faint visual companion, allowing coherent integration times of at least 250 milliseconds. Observations have been made of several visual binaries, including 16 Cyg AB ($m_K = 4.5$ and 4.6) and HD 173648/173649 ($m_K = 4.3$ and ~ 5) to test the performance of the technique. These measurements also demonstrate that phase-referenced visibility measurements can be calibrated at the level of 3–7%.

5.1 Introduction

The sensitivity of a stellar interferometer is limited by the requirement that sufficient photons be collected in a coherence volume ($\tau_0 r_0^2$) to allow an accurate measurement of the fringe phase, and thus to allow fringe tracking. At the Palomar Testbed Interferometer (PTI, Colavita et al. 1999) the atmospheric coherence time (τ_0) in the K band (2.2 μm) is typically 10–20 ms, and the atmospheric coherence diameter (r_0) is ~ 40 cm. This results in an effective tracking limit in dual-star mode (see below) of $m_K \simeq 4.5$, which limits the number of available targets. Phase referencing is a technique intended to improve the limiting magnitude of an interferometer. We report here the first results from testing of phase referencing at PTI.

If a star is too faint to track but appears close in the sky to a brighter star which can be tracked, the usual tracking limit no longer applies (Shao & Colavita 1992b, Quirrenback et

---

[a]The material in this chaper is in press as Lane & Colavita, AJ, 2003.

al. 1994). In this case one can use the measured fringe phase of the bright star to correct, in real time, the fringe phase of the fainter star. This effectively increases the atmospheric coherence time as seen by the second fringe tracker, allowing it to use a longer coherent integration time with a correspondingly fainter tracking limit. For the technique to work well the two stars must be separated by less than an isoplanatic angle, i.e., the angle on the sky over which atmospherically induced motion is well correlated ($\theta_i \propto r_0/h^*$, where h^* is the effective height of the turbulence profile, usually $\theta_i \sim 20$ arc-sec in the K band).

A major use of this technique will be narrow-angle astrometry (Shao & Colavita 1992b, Colavita et al. 1994), which allows one to detect fringes simultaneously on two closely spaced stars, and which can allow astrometric accuracy on the order of tens of micro-arcseconds. Although this type of measurement can be done without the use of phase referencing, the number of suitable target pairs may be quite small (for PTI with two 40 cm apertures: two stars brighter than 4.5 magnitudes and separated by less than 20 arcseconds within the field of regard of the instrument, which results in ~ 4 pairs). However, the situation improves considerably if one can use phase referencing to allow fainter reference stars. Thus phase referencing is required for narrow-angle astrometry to be used on a large scale, particularly in planet searches such as those planned for the Keck Interferometer.

5.2   Instrument Configuration

PTI was designed with an unusual "dual-star" configuration in which the image planes of the apertures can be split (usually by a 50-50 beamsplitter, although a pinhole can be used) such that light can be directed down two different beam paths to two separate beam combiners. Thus it is in effect two independent two-aperture interferometers that share the same apertures. Usually one star (the "primary") is observed on-axis, while the "secondary" star can be anywhere within an annulus with inner radius ~ 8 arcseconds (closer than that and the tip-tilt sensor confuses the primary and secondary stars) and an outer radius of 1 arcminute.

After tip-tilt correction by a fast steering mirror, the starlight from each aperture passes through optical delay lines to correct for geometric and atmospheric optical path-length differences. PTI was designed such that both the primary and secondary starlight beams pass through a common long delay line ("LDL", capable of up to ± 38.3 m of optical delay), after which only the primary beam passes through a short delay line ("SDL", ± 3 cm of optical delay, corresponding to ± 1 arc-minute on the sky). In effect, the primary fringe tracker sees a "primary" delay line with an optical delay given by $\delta_p = \delta_{LDL} + \delta_{SDL}$, while the secondary fringe tracker sees a "secondary" delay line with delay $\delta_s = \delta_{LDL}$. The delay line controller orthogonalizes the commands sent to the physical delay lines such that the fringe trackers can request optical pathlength changes to the primary and secondary delay lines independently. Also, delay modulation (see below) can be applied to either or both of the delay lines.

Optical path-lengths are monitored by several laser metrology gauges, including independent LDL and SDL monitors, as well as a "constant-term" (CT) metrology system, which measures the total difference in optical path delay between the primary and secondary beams throughout the entire optical system out to the apertures. The SDL can use either its local metrology system or the CT to determine position. The latter case provides a way to compensate for piston vibrations of the optics in the starlight path and is used in phase referencing.

PTI is able to operate in three fundamental modes: the simplest is the case when only one fringe tracker operates, tracking and measuring the visibility of a single star. In this mode the 50-50 beamsplitters in the focal planes of the apertures are usually removed, increasing the photon throughput of the instrument. The second observing mode is used for astrometric measurements of similar-magnitude visual binary systems: in this mode both the primary and secondary fringe trackers operate independently with short sample times, and the primary and secondary delay lines are effectively independent. The third observing mode uses phase-referencing, in which the primary fringe tracker tracks a bright ($m_K < 4.5$) star with short sample times, while correcting the measured phase error for both the primary and secondary delay lines. In this mode the secondary fringe tracker can operate with integration times of 100 ms or longer.

5.3 Fringe Tracking & Phase Referencing

Fringe tracking at PTI (Colavita et al. 1999, Colavita 1999) is implemented as follows: the tip-tilt corrected and delay-compensated starlight beams from each aperture are combined at a 50-50 beamsplitter. The output of the beamsplitter is two combined beams, one of which is focused directly onto a single pixel of a NICMOS-3 detector. This channel is usually operated in the astronomical K band $(2.0 - 2.4 \ \mu m)$, and is referred to as the "white-light" channel. The other beam is first spatially filtered by passage through a single-mode fiber and then dispersed with a prism before being focused onto 5–10 pixels (depending on the chosen spectral resolution, typically 65 nm/pixel) on the same detector, and is used as a spectrometer.

The fringe signal is measured by modulating the delay in a sawtooth pattern with an amplitude of one wavelength, and synchronously reading out the detector. For normal operation two sample times are available (10 and 20 ms), while for phase-referenced operation the secondary fringe tracker was modified to allow integration times of 50, 100 and 250 ms. During each sample the detector is first reset, a bias level is read, and then 4 reads are done, one after each quarter-wavelength of modulation. Denoting the integrated intensities in each $\lambda/4$ bin as A, B, C and D, the fringe quadratures are calculated as

$$X = A - C \tag{5.1}$$

$$Y = B - D \tag{5.2}$$

and the total flux as

$$N = A + B + C + D \tag{5.3}$$

After these quantities have been corrected for read-noise and detector biases for each pixel and frame, the fringe visibility is calculated as

$$V^2 = \frac{\pi^2}{2} \frac{X^2 + Y^2}{N^2} \tag{5.4}$$

and the fringe phase is found from

$$\phi = \tan^{-1} \frac{Y}{X} \tag{5.5}$$

This measured phase is "unwrapped" about a Kalman-filter based prediction to provide the phase used by the real-time system.

Once a new phase measurement becomes available, the fringe tracker adjusts the delay line position to keep the fringe phase as close to zero as possible. In practice this is done via an integrating servo (see Appendix A), and as is the case in any servo system, the correction is not perfect. In particular, the fringe tracker cannot control phase errors at frequencies above the servo bandwidth. Given a phase disturbance (the atmosphere) with power spectral density (PSD) $A(f)$, the PSD of the residual phase not corrected by the fringe tracker is given by

$$W(f) = A(f)H_{fb}(f) \qquad (5.6)$$

where $H_{fb}(f)$ is the error (power) rejection of the servo

$$H_{fb}(f) \simeq \frac{1}{1 - 2\frac{f_c}{f}\mathrm{sinc}(\pi f T_s)\sin(2\pi f T_d) + \left(\frac{f_c}{f}\right)^2 \mathrm{sinc}^2(\pi f T_s)} \qquad (5.7)$$

where $\mathrm{sinc}(x) = \sin(x)/x$, f_c is the closed-loop bandwidth of the servo, and T_s and T_d are delays (defined below). For the servo to be stable (not oscillate) the servo gain must be less than unity; typically $f_c \sim 0.1 - 0.21/T_s$. At PTI, $f_c \sim 5$ Hz for 20 ms sample times. T_s is the integration time of the measurement, effectively 15 ms for a sampling time of 20 ms (the lost time is due to modulation retrace and detector reset and settle time). T_d is the effective delay between measurement and correction, including data age. For PTI $T_d = 21.5$ ms for 20 ms sample times. The theoretical servo responses are shown in Fig. 5.1.

Under good seeing conditions the primary fringe tracker can maintain a stable lock on a star for several minutes. Typical uncalibrated visibilities (V^2) in the spectrometer, which includes a spatial filter, are 0.7–0.8 when observing a point source. For the broadband ("white-light") channel, which does not include a spatial filter, typical uncalibrated visibilities are 0.3–0.4.

With a fringe-tracking interferometer such as PTI, the most obvious way to implement phase referencing is simply to apply the delay corrections from the primary fringe tracker to both the primary and secondary delay lines; we call this the "feedback" approach. In this case we expect the power spectrum of the phase seen by the secondary beam combiner to look like the residual phase $W(f)$. Figure 5.2 shows how the unwrapped fringe phase is affected by the atmosphere, and how phase referencing stabilizes it. Fig. 5.3 shows the power spectra of the same two cases, along with the best-fit atmospheric power spectrum. Also shown is the predicted power spectrum, based on application of Eq. 5.7 to the best-fit atmospheric power spectrum.

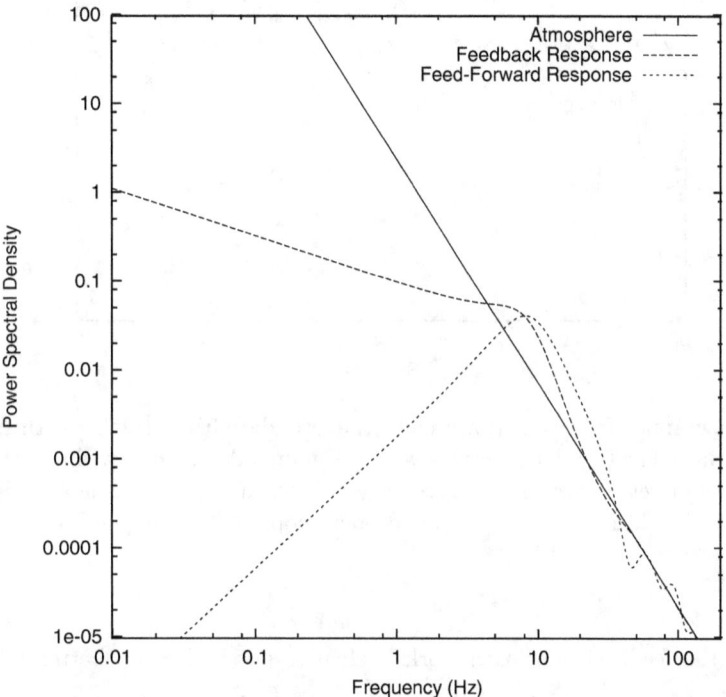

Figure 5.1: Predicted power spectra for phase referencing in the feedback and feedforward cases. The -2.5 power law is the best fit to the atmospheric power spectrum from Fig. 5.3., and the feedback and feedforward theoretical predictions are explained in the text. Model parameters were $f_c = 5$ Hz, $T_d = 21$ ms, $T_s = 15$ ms.

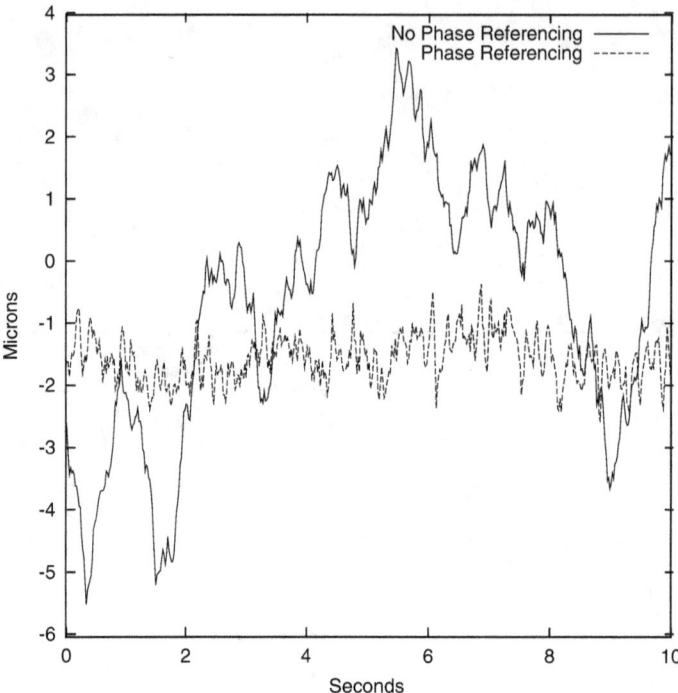

Figure 5.2: Unwrapped fringe position seen by the secondary fringe tracker with and without phase referencing. The two data sections were taken on 4 Aug, 1999, within 200 seconds of each other. The target star was HD 177724 ($m_K = 2.99$, A0V). The secondary fringe tracker was operating with 20 ms sample times and open loop, i.e., measuring but not correcting the phase.

Although the feedback approach works well it is possible to do better. This comes about because the secondary delay line is not in the feedback path of the primary fringe tracker. Therefore there is no issue of servo stability, and the primary fringe tracker can apply all of the measured phase error to the secondary delay line, resulting in improved performance. We refer to this as the "feedforward" approach. In this case the error (power) rejection function is given by the feedback filter function, multiplied by a factor that depends only on the integration time and the time delay between the phase measurement and the application of the correction (see Appendix A for a derivation).

$$H_{ff}(f) = H_{fb}(f) \left(1 - 2\text{sinc}(\pi f T_s)\cos(2\pi f T_{ds}) + \text{sinc}^2(\pi f T_s)\right) \qquad (5.8)$$

The predicted power spectra of the secondary phase, based on the PSD filter functions

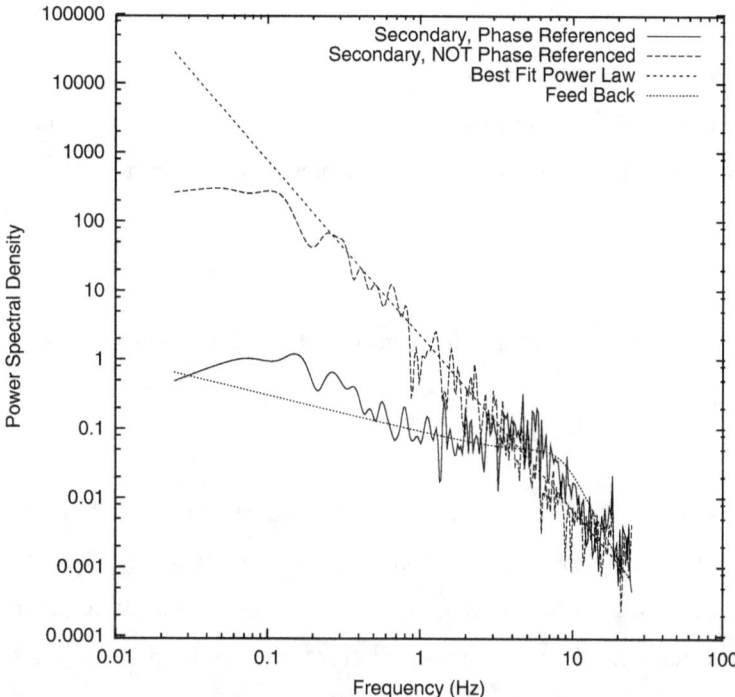

Figure 5.3: Power spectral density of the secondary phase, for both non-phase referenced and phase-referenced data. Note that the non-phase-referenced phase (essentially the atmosphere) is best fit by a power law $A(f) \propto f^{-2.5}$, somewhat shallower than the nominal -8/3 slope of Kolmogorov theory. However, it is similar to the slope seen in other PTI data [Linfield et al. 1999]. Also included is the predicted power spectrum for the phase-referenced case, based on the performance of a feedback servo applied to the best-fit atmospheric power-law (see text).

and the model atmosphere power law of Fig. 5.3, is plotted in Fig. 5.1. As can be seen in the figure, using a feedforward servo the power at low frequencies falls off rapidly, and hence for sufficiently long integration times the feedforward case is expected to result in significantly reduced servo error compared to the feedback case. However, the slight increase in energy near f_c does mean that for short integration times (less than 100 ms) the feedback approach is to be preferred; our results are based on the feedback approach.

5.4 Visibility Reduction

One of the most important performance-limiting factors in phase-referencing is that fluctuations in the fringe position during integration reduce the measured fringe visibility ("smearing" the fringe). This reduction in V^2 can be calculated as (Colavita 1999)

$$V_{sys}^2 = \exp[-(\sigma_\phi)_{hp}^2] \tag{5.9}$$

where $(\sigma_\phi)_{hp}^2$ is the high-pass filtered fluctuation of the phase about the interval mean

$$(\sigma_\phi)_{hp}^2 = \int_0^\infty W(f)[1 - \text{sinc}^2(\pi f T)]df \tag{5.10}$$

where $W(f)$ is the power spectral density of the residual phase and T is the integration time. We calculated the expected reduction in fringe visibility for the theoretical feedback and feedforward servo filter functions applied to the best-fit model atmosphere of Fig. 5.3. The results are shown in Fig. 5.4. We also note that these performance estimates are made based on the response time of the PTI systems; increasing fringe tracker bandwidth can reduce the loss of fringe visibility considerably (Fig. 5.5). However, increasing the fringe tracker bandwidth requires both faster computers and shorter integration times, which in turn necessitates either larger apertures or brighter reference stars.

In addition to adding feedforward, it is possible to tune the fringe tracker by adding a proportional gain term (See Appendix A): the effect of such a term is to reduce the peaking in the response function near the closed-loop frequency, at the expense of increased peaking at higher frequencies. However, given that the atmospheric noise is proportional to $f^{-2.5}$ the net effect is a significant recovery in system visibility (Fig. 5.6).

5.5 Observations

An example of the visibility data produced by the instrument is given in Fig. 5.7. In this experiment the primary fringe tracker continuously tracked 16 Cyg A, providing phase-referencing for the secondary fringe tracker. The secondary FT used a coherent integration time of 100 ms, and switched between observing 16 Cyg A and 16 Cyg B every 130 seconds. In each case we considered only the broadband white-light channel, as the narrowband spec-

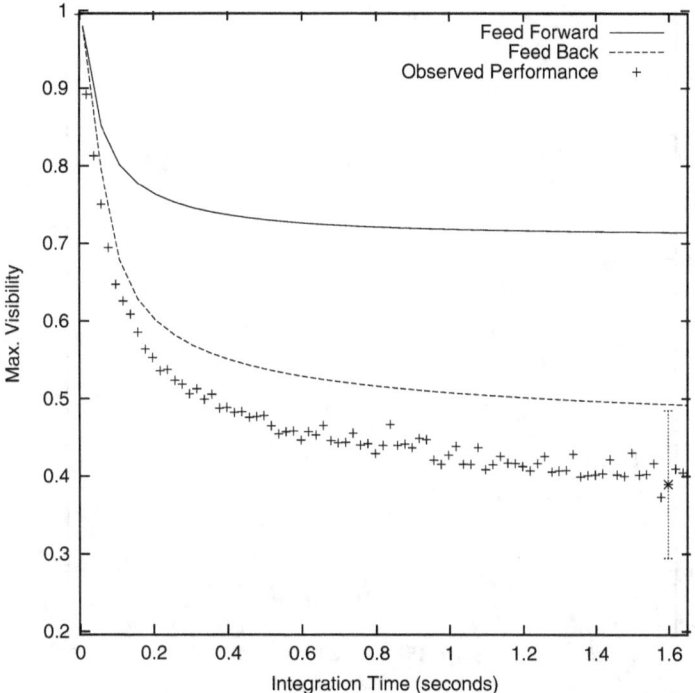

Figure 5.4: Predicted system visibility (V_{sys}^2) as a function of coherent integration time for the feedforward and feedback cases, and the observed visibility reduction obtained by averaging the time-series data shown in Fig. 5.2 Model parameters were $f_c = 5$ Hz, $T_d = 21$ ms, $T_s = 15$ ms.

trometers differed significantly between the two fringe trackers (the primary side includes a spatial filter while the secondary side uses a slit) complicating comparisons of measurement precision. As both 16 Cyg A and B are expected to appear as point sources (the radial velocity companion of 16 Cyg B (Cochran et al. 1997) is far too faint to be directly observed by PTI), we would expect to see a visibility given by (Boden et al. 1998)

$$V^2 = \left[\frac{2 \, J_1(\pi B\theta/\lambda)}{\pi B\theta/\lambda} \right]^2 \tag{5.11}$$

where J_1 is the first-order Bessel function, B is the projected baseline vector magnitude at the star position, θ is the apparent angular diameter of the star, and λ is the center-band wavelength of the interferometer. This model predicts a constant visibility at a level determined by the angular size of the source, wavelength of observation, and projected

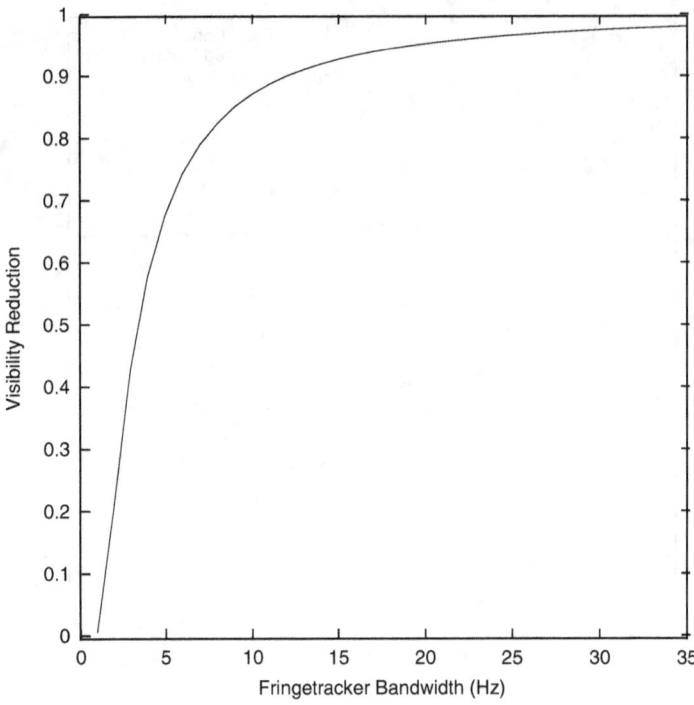

Figure 5.5: Predicted system visibility for a 2-second coherent integration time, as a function of fringe tracker closed-loop bandwidth, assuming a feed-forward servo and the PTI best-fit atmospheric power law. Model parameters were $T_d = 0.1/f_c$ and $T_s = 0.75T_d$.

baseline length of the interferometer.

The scatter (defined as $\sigma_{V^2} = \langle |V_i^2 - \langle V^2 \rangle| \rangle$) around a flat line is $\sim 18\%$ for the uncalibrated phase-referenced data. However, notice that the fluctuations in visibility are common to both the primary (A) and secondary (B) stars – making it possible to calibrate the data by using measurements of star A to calibrate measurements of star B (as done routinely to calibrate non-phase-referenced data). This is done by assuming a uniform-disk model for the star A, and comparing the model visibility of the primary to the observed visibility of that star. From this one can derive the "system visibility" or inherent visibility response of the instrument ($V_{sys}^2 = V_{raw,calibrator}^2/V_{model}^2$). The calibrated data is found by applying $V_{calibrated,target}^2 = V_{raw,target}^2/V_{sys}^2$.

In addition to calibrating the observed fringe visibilities by interleaving observations of target and calibrators, the use of phase referencing means that there are simultaneous observations of the primary star by the "primary" fringe tracker; this makes possible a second

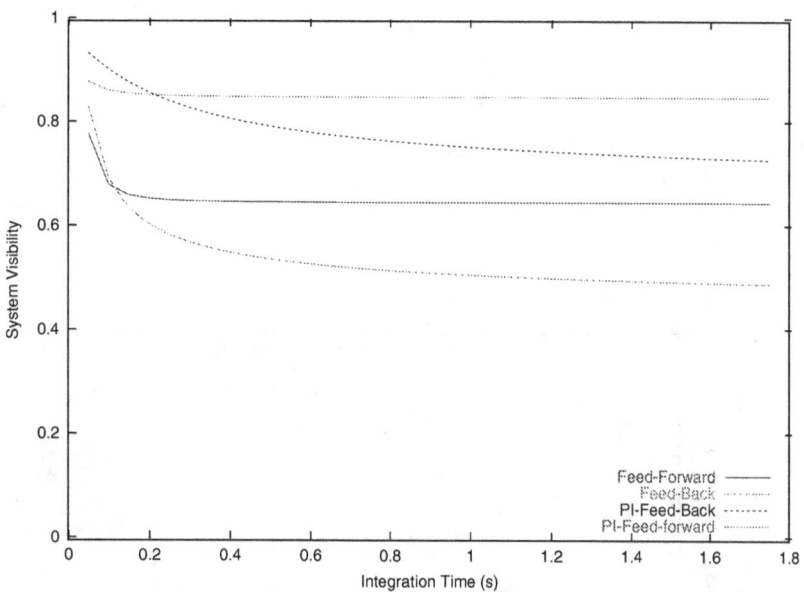

Figure 5.6: This is what the predicted system visibilities are for all 4 cases: regular feedback and feedforward, and the same but with an added proportional gain term. Model parameters were $f_c = 10$ Hz, $T_d = 11$ ms, $T_s = 6.75$ ms, $K_p = 0.5$.

approach to calibrating the data. As can be seen in Fig. 5.5, the fringe visibility measured by the primary fringe tracker is higher than that measured by the secondary fringe tracker, reflecting differences in both integration time and inherent instrumental response (due to a variety of differences in optical quality, alignment, and instrument layout). Nevertheless, it is evident from the figure that short-term changes in system visibility are somewhat correlated (a linear correlation coefficient r = 0.75), as might be expected from the fact that both systems are looking through (nearly) the same atmospheric turbulence. Hence it is possible to use the calibrator visibility measured by the primary fringe tracker to estimate the system visibility (V_{sys}^2). However, in this case it becomes necessary to assume (or measure via other means) an additional time-independent calibration factor, corresponding to the mean ratio of primary and secondary fringe tracker system visibilities.

In order to characterize the precision with which we can calibrate the phase-referenced visibility measurements by either of the above methods, a uniform-disk model was fit to the calibrated data for several sources and integration times (see Table 5.1). The scatter of the measurements around the model, $\sim$ 3–7 %, are fully comparable with non-phase-

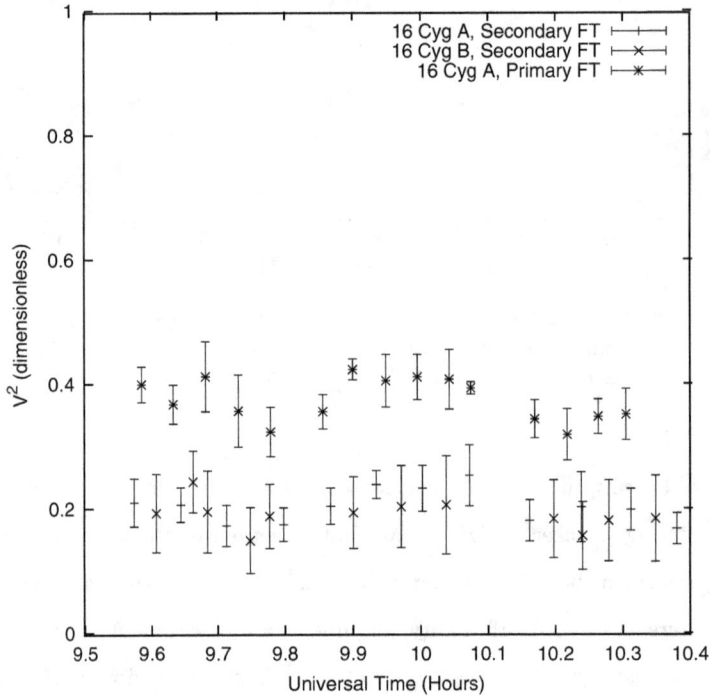

Figure 5.7: Phase-referenced visibilities of the two components of the visual binary 16 Cyg ($m_K = 4.52$, G1.5V, and $m_K = 4.65$, G3V), observed on July 6, 1999 using the broadband fringe-tracker channels. Coherent integration time was 100 ms for the secondary fringe tracker, while the primary fringe tracker used a coherent integration time of 20 ms.

Object	m_K	Date MJD	Coherent Int. Time (ms)	σ_{V^2} Sec. Cal.	σ_{V^2} Pri. Cal.	# Data Points	Estimated Diameter (mas)
61 Cyg B	2.8	51394.31	20	0.051	0.054	49	1.94 ± 0.009
		51365.48	100	0.040	0.034	12	2.14 ± 0.091
		51364.47	250	0.050	0.013	2	1.98 ± 0.319
		51410.31	250	0.025	0.030	8	2.14 ± 0.169
16 Cyg B	4.6	51365.42	100	0.069	0.049	12	0.96 ± 0.35
HD 173649	~ 5	51396.21	250	0.047	0.031	4	0.73 ± 0.22

Table 5.1: Measured scatter of the calibrated visibilities around a uniform-disk model, as measured by the secondary fringe tracker. Two different calibration schemes were tested: "Sec. Cal" refers to the case where visibility measurements were calibrated using interleaved measurements of a calibrator star, measured with the same (secondary) fringe tracker. "Pri. Cal." refers to data calibrated using simultaneous visibility measurements of the calibrator star using a different (primary) fringe tracker. Note that the 20 ms data was not phase-referenced, but is included to provide a comparison of data quality. Each data point corresponds to an incoherent averaging time of 130 seconds. The resulting diameter measurements can be compared to estimates derived from effective temperatures and bolometric fluxes based on archival broad-band photometry: $\theta_{61CygB} = 1.97 \pm 0.03$, $\theta_{16CygB} = 0.551 \pm 0.05$, and $\theta_{HD173649} = 0.3 \pm 0.05$ (all in milli-arcseconds).

referenced performance on much brighter sources. It appears that using the primary fringe tracker as a calibration reference does a comparable job of reducing the scatter in the visibilities. However, this method may be prone to systematic errors in estimating the ratio of system visibilities, and hence we suggest that a hybrid approach that makes use of both types of calibration may be preferable. For instance, one could use the primary fringe tracker to estimate short-term (second to minute timescales) fluctuations in V_{sys}, while using observations of the primary star with the secondary fringe tracker to establish the mean difference in V_{sys} between the two fringe trackers. By making use of the information provided by the primary fringe tracker in this manner one can reduce the number of calibration observations required.

5.6   Conclusion

By synthetically increasing the apparent atmospheric coherence time, phase-referencing promises a dramatic increase in the sensitivity of a stellar interferometer. Initial results from phase-referencing experiments at PTI are encouraging, and demonstrate that it is possible to increase coherent integration times by at least a factor of 10. We also demonstrate that phase-referenced measurements of source visibilities can be calibrated to at least the $3 - 7\%$

level, depending on source brightness and observing conditions. This is similar to what can be done with non-phase-referenced PTI data, and hence we see no loss of precision in using phase-referencing. Note that these data were obtained without the use of a spatial filter; adding one should improve measurement precision. Future experiments will be conducted to determine the ultimate improvement possible, as well as the effect of anisoplanatism.

Chapter 6

Adaptive Optics Observations of the Binary Brown Dwarf GJ 569B

We present photometric, astrometric and spectroscopic observations of the nearby (9.8 pc) low-mass binary Gl 569Bab (in turn being a companion to the early-M star Gl 569A), made with the Keck adaptive optics facility. Having observed Gl 569Bab since August 1999, we are able to see orbital motion and to determine the orbital parameters of the pair. We find the orbital period to be 876 ± 9 days, the semi-major axis to be 0.90 ± 0.01 AU, the eccentricity to be 0.32 ± 0.01 and the inclination of the system to be 34 ± 3 degrees ($1\text{-}\sigma$). The total mass is found to be $0.125^{+0.027}_{-0.022} M_\odot$ ($3\text{-}\sigma$). In addition, we have obtained low resolution ($R = 1500$–1700) near-infrared spectra of each of the components in the J- and K-bands. We determine the spectral types of the objects to be M8.5V (Gl 569Ba) and M9V (Gl 569Bb) with an uncertainty of half a subclass. We also present new J- and K-band photometry which allows us to accurately place the objects in the HR diagram. Most likely the binary system is comprised of two brown dwarfs with a mass ratio of 0.89 and with an age of approximately 300 Myr.

6.1 Introduction

Brown dwarfs (BDs), despite sometimes being labeled "failed stars", are very interesting objects. According to our current understanding, BDs may represent the extreme low-mass end of star formation in which the mass is too small to sustain thermonuclear fusion. This stellar-substellar transition (also defined as the minimum mass at which the internal energy provided by nuclear burning quickly balances the gravitational contraction energy) is

---

[a]The material in this chaper was previously published as Lane et al., ApJ, 560, 390–399 ,2001.

expected to occur around 0.075 $M_\odot$ (Baraffe et al. 1998) for objects with solar metallicities. This value is slightly lower in recent models, which predict a transition mass of 0.072 $M_\odot$ (Chabrier & Baraffe 2000). Objects with masses below this limit never reach the stable hydrogen burning main sequence, but instead cool down as they age (Burrows et al. 1997; Chabrier et al. 2000) so that their surface temperatures and luminosities strongly depend on age as well as mass. As these cooling curves rely on poorly tested theoretical models, it is highly desirable to calibrate them with direct measurements.

Obtaining dynamical masses for very low-mass (VLM, $M \leq 0.2\,M_\odot$) stars and BDs from binaries is a challenging prospect, as so far there are no known eclipsing binary VLM stars or BDs. However, there are a number of known wide, non-eclipsing VLM and BD binaries (Kirkpatrick et al. 2001; Gizis, Kirkpatrick & Wilson 2001; Leinert et al. 2001; see also Reid et al. 2001a and references therein) that have been observed with a range of instruments and techniques, including ground-based infrared and optical imaging, speckle interferometry, adaptive optics (AO), and the Hubble Space Telescope. These binary systems promise to yield highly accurate dynamical masses, although they tend to have long periods and hence will require patience.

In the absence of direct mass measurements for VLM/BD objects, one has to rely on indirect methods that may constrain mass ranges but that do not provide high-precision mass values. One very useful such technique is the lithium test (Magazzù, Martín & Rebolo 1993). Lithium is an element that is easily destroyed under the conditions prevalent in stellar interiors at temperatures slightly below those required for hydrogen burning. Objects more massive than $\sim 0.060\,M_\odot$ have their primordial lithium abundances depleted as long as they are fully convective. The lithium test has been used to confirm BD candidates in the Pleiades (e.g., Basri, Marcy & Graham 1996) and in the field (e.g., Martín, Basri & Zapatero Osorio 1999; Kirkpatrick et al. 1999; Tinney, Delfosse & Forveille 1997).

Gl 569A is a nearby ($d = 9.8\,\mathrm{pc}$), chromospherically active late-type (M2.5V) star. Forrest, Skrutskie & Shure (1988) first reported a possible BD companion with a separation of 5 arcseconds from the primary. Based on a low resolution spectrum Henry & Kirkpatrick (1990) classified this object as an M8.5 dwarf with a mass of $0.09 \pm 0.02 M_\odot$. More recently, Martín et al. (2000a) resolved the companion into two separate objects (Gl 569Ba and Bb) using AO observations with the Keck II telescope. They estimated the orbital period of the Ba-Bb binary to be around 3 years, and the total mass of the binary pair to be in the range $0.09 - 0.15\,M_\odot$. Herein we present the results of extensive follow-up observations of this interesting pair, including improved photometry as well as near-IR spectroscopy and astrometry of the resolved binary components. We use the photometry to accurately place the objects in the Hertzsprung-Russell (HR) diagram, the spectroscopy to derive spectral types, and the astrometry to derive the total mass of the system from its orbital parameters.

Object	J	K	$J - K$	$\log L/L_\odot$	T_{eff} (K)	$M/M_\odot$
Gl 569Bab	10.61 ± 0.05	9.45 ± 0.05	1.16 ± 0.07	-3.17 ± 0.07		0.101–0.150
Gl 569Ba	11.14 ± 0.07	10.02 ± 0.08	1.12 ± 0.10	-3.39 ± 0.07	2440 ± 100	0.055–0.078
Gl 569Bb	11.65 ± 0.07	10.43 ± 0.08	1.22 ± 0.10	-3.56 ± 0.07	2305 ± 100	0.048–0.070

Table 6.1: Photometry (CIT system) and physical parameters of Gl 569Ba and Gl 569Bb The relative photometry between Gl 569Ba and Gl 569Bb is known to a better accuracy (see text).

6.2 NIR Photometry and Astrometry

6.2.1 The composite pair: Gl 569Bab

Broad-band near-infrared photometry of the composite pair Gl 569Bab is available in the literature (Forrest et al. 1988; Becklin & Zuckerman 1988). However, the measurement uncertainties claimed by the authors are too large for an accurate placement of these objects in the HR diagram or for direct comparison with theoretical evolutionary models. With the objective of improving the photometric data, we have collected J and K_{short} direct images of the system Gl 569A and Gl 569Bab with the near-infrared camera (Hg Cd Te detector, 256×256 elements) mounted at the Cassegrain focus of the 1.5-m Carlos Sánchez Telescope (CST, Teide Observatory) on February 8, 2001. The observations were performed through the "narrow-optics" of the instrument, which provides a pixel projection of 0.4 arcseconds onto the sky. The atmospheric seeing conditions during the night of the observations were fairly stable around 1 arcseconds, which allowed us to easily separate the M2.5-type star from the pair Gl 569Bab. This latter object was not resolved into its two components. The total integration times were 5 s and 40 s in J and K_{short} filters, respectively. A five-position dither pattern was used to obtain the images; each image consisted of 4 (J) or 8 (K) co-added exposures of 0.25 s (J) and 0.5 s (K) respectively. The dither pattern was repeated twice for the K-band observations.

Dithered images were combined in order to obtain the sky background, which was later substracted from each single frame. Gl 569Bab is clearly detected in individual images, and we have obtained aperture photometry on each of them using PHOT in IRAF.[a] Instrumental magnitudes were placed on the UKIRT photometric system using observations of the standard star HD 136754 (Casali & Hawarden 1992), which was imaged with the same instrumental configuration just before and immediately after our target. Both the science target and the standard star were observed at similar air masses. The photometric error of the calibration was ± 0.03 mag in both filters. K_{short} displays a different bandpass compared to K_{UKIRT}; the transformation between these two filters is not well defined yet, albeit for objects as red as Gl 569Bab it has been estimated at $K_{short} - K_{UKIRT} = 0.035$ (Hodgkin et al. 1999, and references therein). We have applied this correction to our photometry as well as the relations given in Leggett, Allard & Hauschildt (1998) to convert UKIRT data into the CIT photometric system. The final average magnitudes derived for Gl 569Bab are given in Table 6.1, where the photometric errors listed correspond to typical $1\,\sigma$ uncertainties of single measurements.

The astrometry of Gl 569Bab relative to the bright primary Gl 569A as measured on the CST data (MJD = 51948.202) is the following: angular separation of 4.arcsec 890 ± 0.arcsec 040, and position angle of $30° \pm 3°$. We note that these values differ from those published in Forrest et al. (1988) by more than $2\,\sigma$, providing evidence of the orbital motion of Gl 569Bab around the M2.5-type star in the time interval of roughly 15 years.

6.2.2   Adaptive Optics Imaging of Gl 569Ba-Bb pair

Gl 569Bab was observed on 9 occasions between August 1999 and September 2001 with the Keck II AO system (Wizinowich et al. 1988). The first 3 observations made use of the KCAM camera with a NICMOS-3 infrared array. The later observations made use of the slit-viewing camera (SCAM) associated with the NIRSPEC instrument (McLean et al. 1998). SCAM uses a PICNIC Hg Cd Te array. For both cameras the pixel scale was 0.0175 arcseconds and the field of view was 4.48 × 4.48 arcsec, except for the May 2001 observation when the SCAM pixel scale was changed to 0.0168 arcsec. Exposures were generally obtained

---

[a]IRAF is distributed by National Optical Astronomy Observatories, which is operated by the Association of Universities for Research in Astronomy, Inc., under contract with the National Science Foundation.

in the K'-band, except for 2000 Febuary when the observations were taken in the J-band. Exact exposure times varied, but typically consisted of 30 coadded 2-second exposures. We used a 1% transmission neutral density filter in the beam to prevent saturating the bright primary star. Flat-field correction was performed using twilight exposures, while sky subtraction made use of images of an adjacent field observed immediately after Gl 569B. Corrected seeing varied between 0.05 arcsec and 0.08 arcsec. For three of the observations the primary star (Gl 569A) was also in the field of view, providing an in-field astrometric reference. For the other observations the primary was either not observed or was saturated. No photometric standard stars were observed in any of the epochs, so we cannot provide absolute photometric calibrations for the AO observations. Figure 6.1 shows the resulting images of the pair Gl 569Bab at six different epochs. This figure clearly demonstrates that this system is resolved into a binary and that orbital motion is evident.

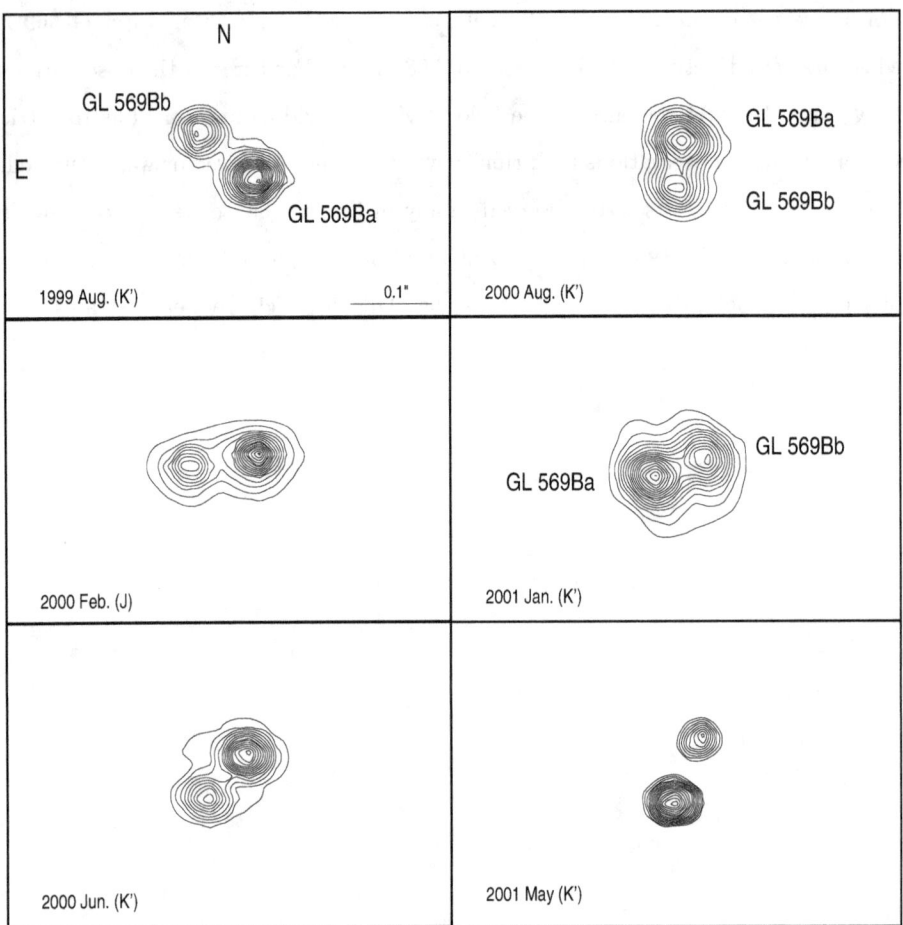

Figure 6.1: Contour images of the Gl569Ba-Bb pair (near-IR filter is given in brackets) showing the orbital motion. These data have been obtained with the Adaptive Optics facility of the Keck II telescope and with the KCAM (first 2 epochs) and SCAM/NIRSPEC (last 7 epochs) instruments. Data of August 1999, when the binary was resolved for the first time, were presented in Martín et al. (2000a).

Date	Epoch (MJD)	Separation (arcsec)	P.A. (°)
1999 Aug 29	51419.270	0.101 ± 0.001	46.8 ± 3
2000 Feb 18	51592.600	0.092 ± 0.001	98.2 ± 3
2000 Feb 25	51599.644	0.090 ± 0.001	100.4 ± 2
2000 Jun 20	51715.408	0.076 ± 0.003	138.6 ± 2
2000 Aug 24	51780.283	0.059 ± 0.001	178.4 ± 2
2001 Jan 09	51918.665	0.073 ± 0.002	291.4 ± 2
2001 May 10	52039.410	0.097 ± 0.001	341.1 ± 3
2001 Jun 27	52088.291	0.102 ± 0.001	352.6 ± 2
2001 Aug 31	52153.221	0.103 ± 0.001	9.7 ± 2

Table 6.2: Astrometry of Gl 569Ba-Bb.

We used the DAOPHOT package (in IRAF) for data reduction and analysis. The point spread functions of the objects in each frame were fitted with an elliptical Gaussian function, providing relative astrometry (Table 6.2) and photometry. The relative photometry of Gl 569Ba and Gl 569Bb was derived by computing the ratio of the amplitudes of the best fitting Gaussians (Table 6.1). Gl 569Bb is fainter by 0.51 ± 0.02 mag and 0.41 ± 0.03 mag in the J- and K-bands, respectively. This makes this object redder in $(J - K)$ by 0.10 ± 0.04 mag. The error bars take into account the dispersion observed from image to image, and from one observing run to another. We do not find a significant relative photometric variability in any member of the pair within $3\,\sigma$ the uncertainties. With the relative brightness of the two components and the combined flux known, it is possible to derive the individual absolute magnitudes of Gl 569Ba and Gl 569Bb. We list in Table 6.1 the resulting decomposition for the J- and K-bands. The corresponding error bars incorporate the photometric uncertainties of the combined system Gl 569Bab and the uncertainties of the relative photometry. We are confident that the latter is determined with a higher accuracy.

6.3 Low-Resolution NIR Spectra

We have obtained low-resolution spectra of Gl 569Ba and Gl 569Bb in the J- (1.158–1.368 μm) and K-bands (1.992–2.420 μm) using the cross-dispersion spectrograph NIRSPEC and the AO facility at the Keck II telescope. The data were collected on June 20, 2000. The raw seeing and transparency conditions were very good during the observations, and the AO correction applied to the primary star Gl 569A provided well resolved images of the binary system Gl 569Bab (AO corrected seeing of 0.05 arcsec). NIRSPEC in its spectroscopic mode is equipped with an Aladdin InSb 1024×1024 detector with a pixel projecting 0.0185 arcsec onto the sky. For the present study, we selected the low resolution spectroscopic mode which provides nominal dispersions of 2.8 Å/pix and 4.2 Å/pix in the J- and K-bands, respectively. The 3 pixel-wide slit was aligned with the two components of the binary (PA ~ 139 deg) so that both targets were observed simultaneously.

Total exposure times were 240 sec and 400 sec for the J-band and K-band spectra, respectively. The observing strategy employed was as follows: 6 (J) and 10 (K) individual integrations of 20 sec (J) and 10 sec (K) each at two different positions along the entrance slit separated by about 1.8 arcsec. This procedure was repeated twice in the K-band. In order to remove telluric absorptions due to the Earth's atmosphere, the near-infrared featureless A0V-type star HR 5567 was observed very close in time and in air mass (within 0.05 air masses). Calibration images (argon arc lamp emission spectra and white-light spectra) were systematically taken after observing each source.

Raw data were reduced following conventional techniques in the near-infrared. Nodded images were subtracted to remove the sky background and dark current. The spectra of the sources and of the calibration lamps were then extracted using subroutines of the TWODSPEC package available in IRAF. The extraction apertures of Gl 569Ba and Gl 569Bb were selected so that cross-contamination was less than 10%. The extracted spectra of the sources were divided by their corresponding normalized extracted flat-fields, and calibrated in wavelength. The 1σ dispersion of the fourth-order polynomial fit was 0.4 Å and 1.0 Å in the J and K spectra, respectively. The hydrogen Pβ absorption line at 1.2818 μm and the Bγ absorption line at 2.1655 μm in the spectra of HR 5567 were interpolated before they were used for division into the corresponding science spectra. We are confident that the science spectra have good cancellation of atmospheric features. To complete the data reduction, we multiplied the spectra of our targets by the black body spectrum for the temperature of 9480 K, which corresponds to the A0V class (Allen 2000).

6.3.1   Spectral Types, Atomic and Molecular Features

The resultant average spectra with a resolution of $R = 1500$ in J and $R = 1700$ in K are depicted in Fig. 6.3. The strongest molecular and atomic features are indicated following

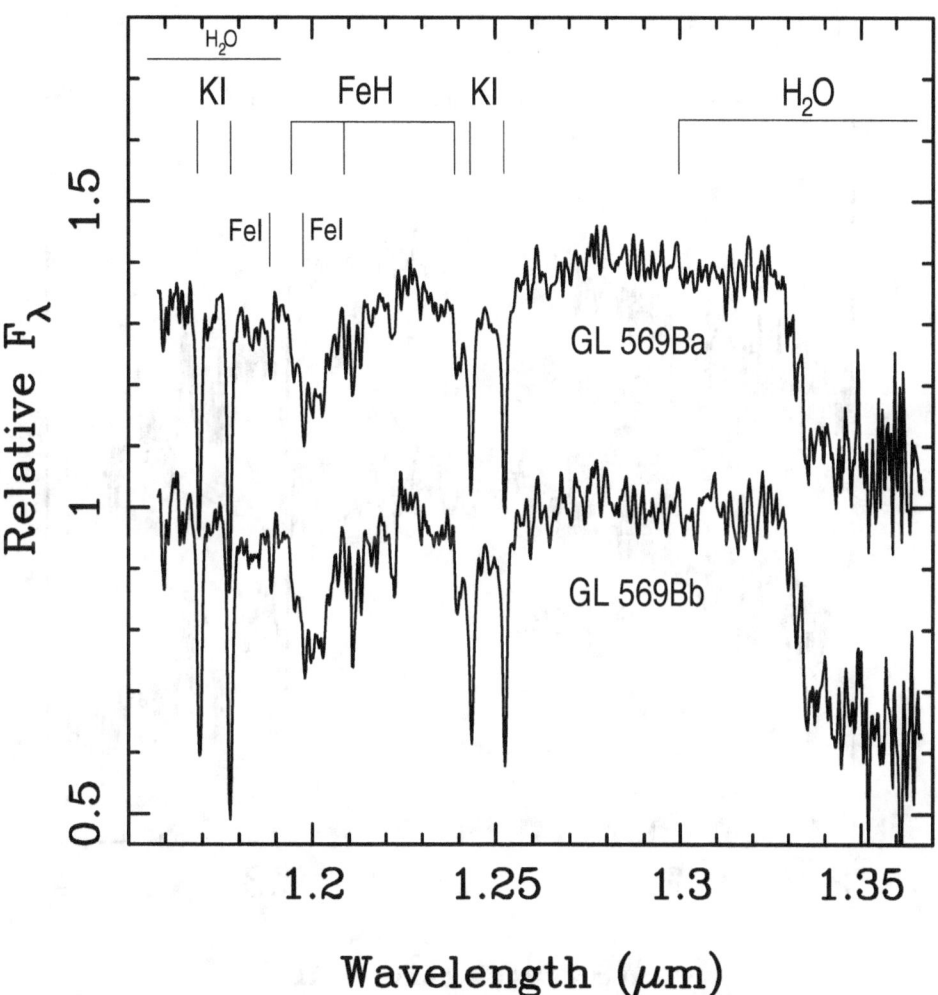

Figure 6.2: *J*-band NIRSPEC spectra of Gl 569Ba and Gl 569Bb obtained using the AO system of the Keck II telescope. Some features have been identified after Jones et al. (1994) and McLean et al. (2000). The spectra have been normalized to unity at $1.29\,\mu$m and at $2.19\,\mu$m. An offset has been added to Gl 569Ba's data for clarity.

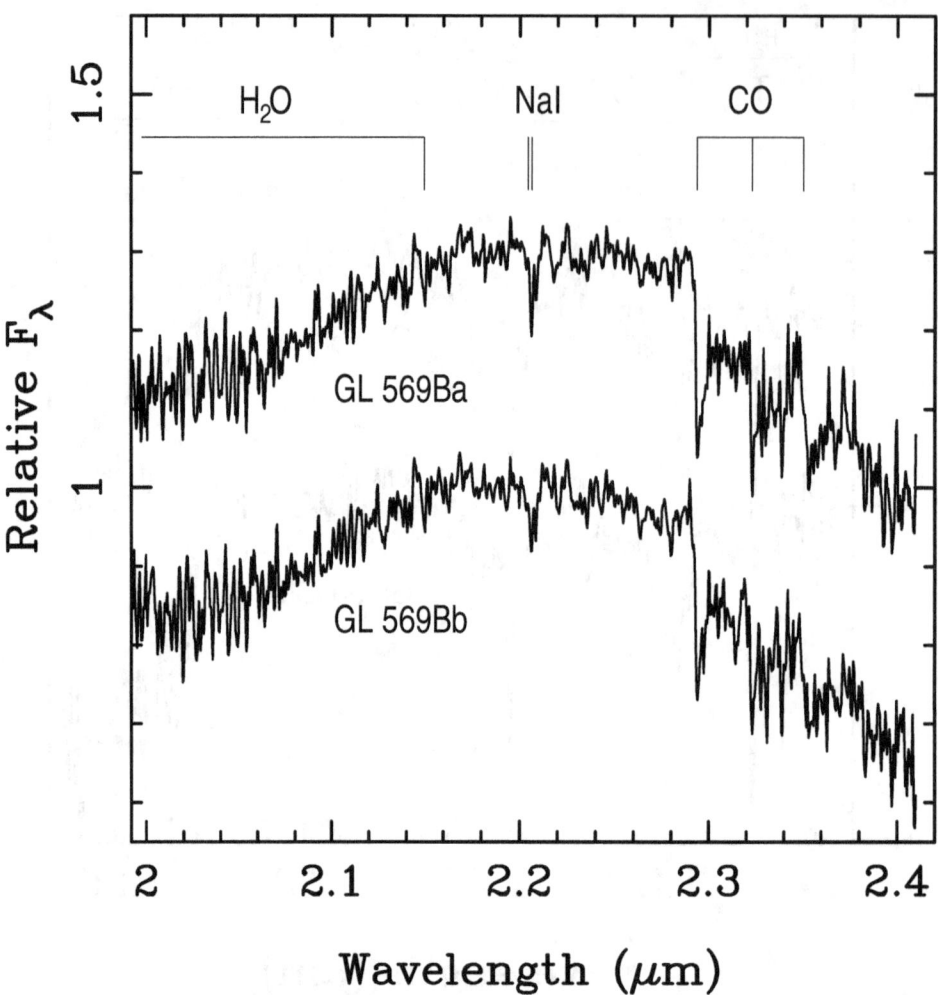

Figure 6.3: *K*-band NIRSPEC spectra of Gl 569Ba and Gl 569Bb obtained using the AO system of the Keck II telescope. Some features have been identified after Jones et al. (1994) and McLean et al. (2000). The spectra have been normalized to unity at 1.29 μm and at 2.19 μm. An offset has been added to Gl 569Ba's data for clarity.

Object	SpT	K I wavelength (μm)				Na I wavelength (μm)		H_2O^a	CO^b
		1.169^c	1.177^c	1.244^c	1.253^c	2.206^c	2.209^c	1.330^c	2.294^c
Gl 569Ba	M8.5	6.5	7.5	4.7	4.8	1.4	0.70	0.72	1.21
Gl 569Bb	M9.0	6.5	7.7	5.2	5.1	0.9	0.50	0.70	1.24

Table 6.3: K I and Na I equivalent widths (Å) and the strengths of the H_2O and CO bands. Uncertainties are ± 0.5 for the spectral classification, 10% for equivalent widths and 5% for the flux ratios.

the identifications provided by Jones et al. (1996) and McLean et al. (2000). The spectra of both components, Gl 569Ba and Gl 569Bb, are indeed very similar. The composite spectrum of Gl 569Bab in the optical has been previously studied by Henry & Kirkpatrick (1990) and Kirkpatrick, Henry & McCarthy (1991), who derived a dwarf spectral type of M8.5. In addition, this object is listed in the Table 1 of Kirkpatrick et al. (1991) as a primary dwarf spectral standard. Our data agree with this measurement for the bright component Gl 569Ba, and also provide evidence that Gl 569Bb is not significantly cooler. This is fully consistent with the photometry presented above.

We have obtained equivalent widths of the strongest observed atomic absorptions of K I and Na I in the spectra; the measurements are given in Table 6.3. Due to the low resolution of our data, the majority of these lines are considerably blended with other spectral features, e.g. the K I line at $1.2435\,\mu$m is contaminated by a strong molecular band of FeH. The values in Table 6.3 have been extracted adopting the base of each line as the continuum. We find typical standard deviations in equivalent width close to 10% over the reasonable range of possible continua. Although this procedure does not give an absolute equivalent width, it is commonly used by different authors, and allows us to compare our values with those published in the literature. We have also measured the strengths of the H_2O band at $1.330\,\mu$m and the CO band at $2.294\,\mu$m in a way similar to that described in McLean et al. (2000), Reid et al. (2001b) and Jones et al. (1994). Our measurements are listed in Table 6.3 with uncertainties of about 5%. All these values are comparable to those obtained from similar spectral type field stars, which suggests that neither Gl 569Ba nor Gl 569Bb have very discrepant metallicities or gravity.

[a]Ratio of the average flux in a $0.02\,\mu$m window centred on $1.34\,\mu$m and on $1.29\,\mu$m (Reid et al. 2001a).
[b]Ratio of the average flux in a $0.06\,\mu$m window centred on $2.25\,\mu$m and on $2.33\,\mu$m (Jones et al. 1994).
[c]All wavelengths in μm.

We note that the equivalent widths of the K I lines and the H$_2$O and CO absorptions in Gl 569Bb appear to be slightly larger than in Gl 569Ba, while the Na I lines in the K-band spectrum are smaller. This trend is observed for decreasing temperatures (Jones et al. 1994), and clearly indicates the cooler nature of Gl 569Bb. By fitting a polynomial spectral type-equivalent width relation to the data available for spectral standard stars (see Reid et al. 2001b), we conclude that the differences between Ba & Bb in our measurements are consistent with Gl 569Bb being half a subclass cooler. This would make Gl 569Bb an M9-dwarf (± 0.5 subclasses).

6.3.2 Radial Velocities

We used our low resolution near-IR spectra taken on June 20, 2000 (MJD $=51715.365$) to compute the relative radial velocity of Gl 569Bb and Gl 569Ba via Fourier cross-correlation. Because the spectroscopic data have been corrected for telluric lines, we do not expect these lines to be a large source of uncertainty. Unfortunately, no spectra were taken of the primary star Gl 569A, so we cannot determine the relative radial velocity of the pair with respect it. The velocity dispersion of the data is rather poor ($1\,\mathrm{pixel} \sim 66\,\mathrm{km\,s^{-1}}$ in J and $\sim 57\,\mathrm{km\,s^{-1}}$ in K). Nevertheless, the cross-correlation technique was able to achieve precisions of about $1/4$ pixel, so we obtained a relative radial velocity accurate to about $15\,\mathrm{km\,s^{-1}}$. We verified this by cross-correlating individual spectra of each component against itself. The relative velocity (Gl 569Bb cross-correlated with Gl 569Ba) we measure is $25\,\mathrm{km\,s^{-1}}$ in J and $6\,\mathrm{km\,s^{-1}}$ in K, with an average value of $15.5\,\mathrm{km\,s^{-1}}$. The peak-to-peak radial velocity variation of the system on the basis of the orbital solution presented in next section is around $14\,\mathrm{km\,s^{-1}}$; our measurement is consistent within the error bar with the expected value at the epoch of the observations. However, this error bar is rather large and prevents us from making further analysis (like the presence of invisible companions). The maximum peak of the cross-correlation function is in the range 0.93–0.97, which indicates the similarity and the high signal-to-noise ratio of the spectra.

6.4   Orbit Determination and Total Mass of the Pair

We determined the apparent orbit of the binary pair Gl 569Bab by fitting a Keplerian model to the relative astrometric data shown in Table 6.2. As the fit is non-linear in the orbital

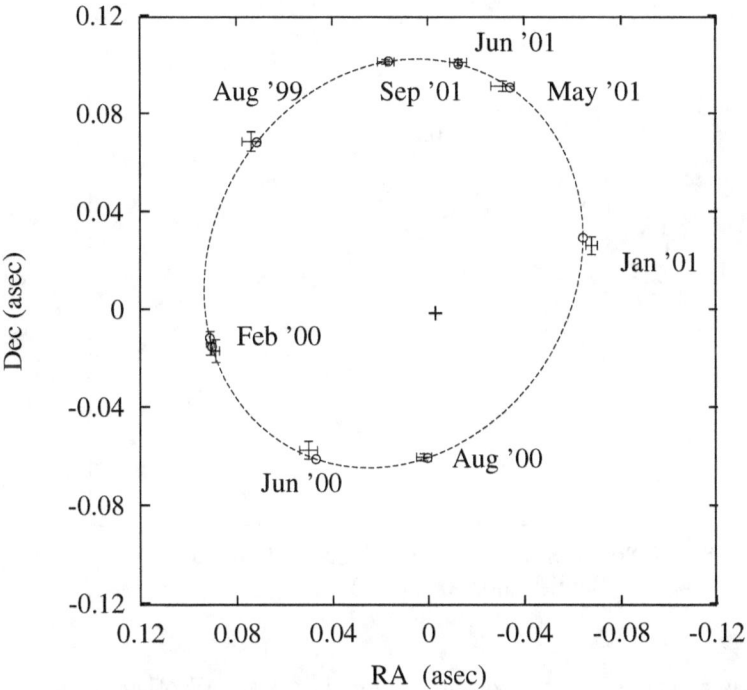

Figure 6.4: The relative astrometry of the Gl 569Ba-Bb pair, together with the best-fit orbit (dotted ellipse). Error-bar crosses denote measurements and circles indicate the predicted location on the orbit at the time of the observations. North is up and East is to the left.

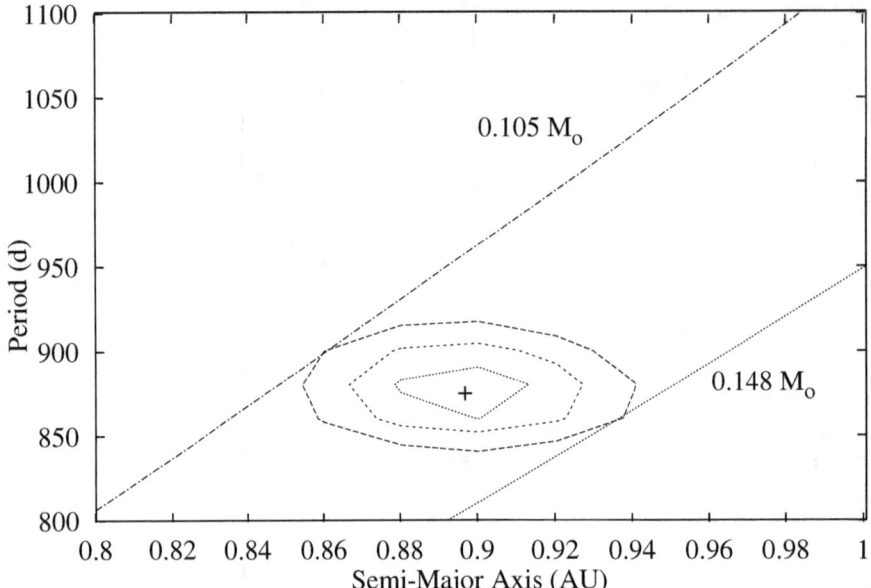

Figure 6.5: The χ^2 as a function of period and semi-major axis. All other parameters are selected to provide the lowest χ^2. The contours give the $1\,\sigma$, $2\,\sigma$ and $3\,\sigma$ uncertainties in the two parameters. The two diagonal lines correspond to combinations of period and semi-major axis giving a total mass of $0.148\,M_\odot$ and $0.105\,M_\odot$, respectively. These are the upper and lower limits that we have adopted for the total mass of the binary Gl 569Bab. The cross indicates the preferred solution.

elements, we made use of a gradient-following fitting routine (Press 1992) to find the optimal (in a chi-squared, χ^2, sense) orbital parameter set. As there may be many local minima in the χ^2 manifold, and our gradient following routine moves strictly downhill, we ensured that it found the global minimum by starting it at a range of different locations in parameter space. The best-fit parameters are given in Table 6.4, and the visual orbit is shown together with the astrometry in Fig. 6.4. The rms of the residuals is 0.0024 arcsec, and the residuals do not show any long-term drifts. Uncertainties (assuming normal errors) were estimated from the covariance matrix of the fit, scaled by the reduced chi-squared (0.96). As the uncertainties in semi-major axis and period may be correlated, in Fig. 6.5 we plot the χ^2 as a function of those two parameters. The mass uncertainty was estimated by finding contours in parameter space where the χ^2 was increased by 2.3, 6.2 and 11.8 respectively, corresponding to the $1\,\sigma$, $2\,\sigma$ and $3\,\sigma$ contours. The $1\,\sigma$, $2\,\sigma$ and $3\,\sigma$ mass ranges are 0.114–0.135, 0.107–0.142 and 0.101–0.150 $M_\odot$ respectively, while the best-fit values for the period and semi-major axis correspond to a total mass of 0.125 $M_\odot$. From the relative photometry we infer that the mass ratio of the binary is close to, but not exactly, equal. Hence the $2\,\sigma$ upper mass limit of the secondary Gl 569Bb is less than 0.071 $M_\odot$, i.e. below the hydrogen burning mass threshold, making it a likely brown dwarf.

It is important to note that although we were able to obtain reliable relative astrometry of the Ba-Bb pair (over a separation of ~ 0.1 arcsec), the uncertainties in plate scale and orientation and the saturation of the bright star in some images were such that we were unable to reliably measure the orbital motion of each component of the Ba-Bb pair with respect to a separate reference, i.e. Gl 569A (located ~ 5 arcsec away). Hence, while we can determine the relative orbit of the Ba-Bb pair to a high degree of precision, we cannot astrometrically determine the mass ratio of the two components. Further studies are needed to confirm or discard the substellar nature of the primary Gl 569Ba. We will combine our photometry (absolute and relative), the total mass of the system and additional information available in the literature to compare the binary with the most recent evolutionary models.

Parameter	Value
Period, P	876 $\pm$ 9 day
Eccentricity, e	0.32 $\pm$ 0.01
Semi-Major Axis, a	0.90 $\pm$ 0.01 AU
Inclination, i	34 $\pm$ 3 deg
Arg. Periapsis, ω	77 $\pm$ 2 deg
Long. of Ascending Node, Ω	142 $\pm$ 2 deg
Epoch (MJD), T	51821.8 $\pm$ 3 day

Table 6.4: Orbital parameters of Gl 569Ba-Bb. Uncertainties are 1-σ. Note that the toal mass uncertainty is smaller than this indicates, see Fig. 6.5 and text.

6.5   Discussion

6.5.1   Color-Magnitude Diagram

Figure 6.6 depicts the location of the Gl569Bab pair in the near-infrared color-magnitude diagram. To convert their observed magnitudes into absolute magnitudes we have used the astrometric parallax provided by Hipparcos (0.arcsec 10191 $\pm$ 0. arcsec 00167, Perryman et al. 1997), which is very similar to previous astrometric measurements (Heintz 1991). Also shown in this figure are the locations of very late-type dwarfs in the Pleiades cluster ($\sim$120 Myr, Basri et al. 1996; Martín et al. 1998; Stauffer, Schultz & Kirkpatrick 1998, we use a distance of 120 pc) which have photometry available in the literature (Festin 1998; Martín et al. 2000b), and of objects in the field. Absolute magnitudes and colors of M-type field standard stars have been taken from tables published in Kirkpatrick & McCarthy (1994) and Leggett et al. (1998). For L-type field dwarfs we have adopted the average near-infrared colors provided in Kirkpatrick et al. (2000), and have averaged absolute K magnitudes for those objects with parallax available in the literature (see Reid et al. 2000, 2001b; Kirkpatrick et al. 2000).

Overplotted onto the observed data in Fig. 6.6 are the 0.5, 1.0 and 5.0 Gyr theoretical solar composition isochrones from the evolutionary models of the Lyon group (Chabrier et al. 2000; Baraffe et al. 1998) and from those of the Arizona group (Burrows et al. 1997). We have adopted solar metallicity in our studies because the photospheric abundance of the bright star Gl569A has been determined to be very close to solar ([Fe/H] = –0.15, Zboril & Byrne 1998). Although the Lyon models do provide magnitudes and colors in

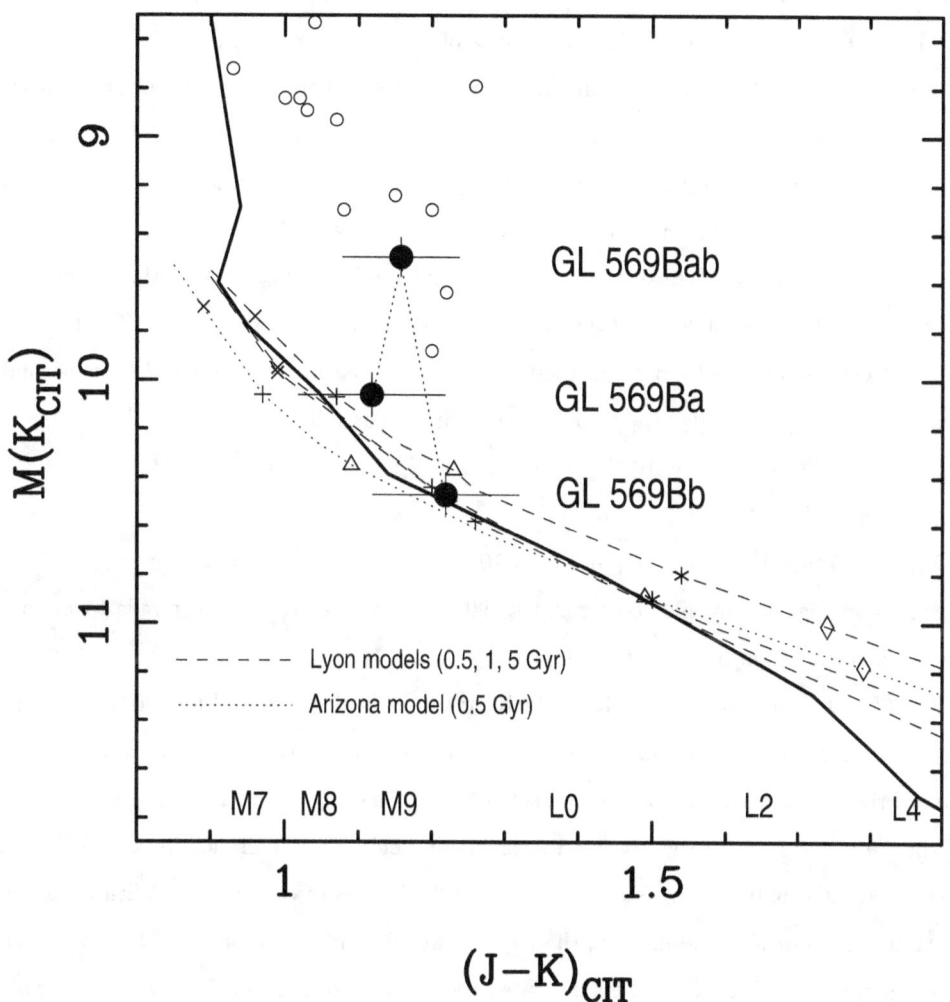

Figure 6.6: Infrared color-magnitude diagram displaying the location of Gl 569Bab (combined light) and each of the two components resolved with the Keck II Adaptive Optics system (filled circles). Pleiades members (~120 Myr) are plotted with open circles, and the location delineated by field dwarfs with known parallax is shown with a thick full line. Isochrones provided by the Lyon group (0.5, 1.0 and 5.0 Gyr, Chabrier et al. 2000 — dashed lines) and by the Arizona group (0.5 Gyr, Burrows et al. 1997 — dotted line) are also overplotted in the diagram. Masses of 0.09 $M_\odot$ (crosses), 0.08 $M_\odot$ (plus-signs), 0.072 $M_\odot$ (open triangles), 0.060 $M_\odot$ (asterisks) and 0.055 $M_\odot$ (diamonds) are marked with crosses on the isochrones. We indicate spectral types as a function of the $(J - K)$ color on the bottom of the figure.

the filters of interest, we preferred to compute them from the predicted luminosity and effective temperature at a given mass and age. This allows a direct comparison of the two sets of interior models and minimizes the effects of possible errors in the model atmosphere synthesis. We converted effective temperatures (T_{eff}) into colors by using the temperature scales of Leggett et al. (1998) for M-dwarfs and of Basri et al. (2000) for late-M and L-dwarfs. These two temperature scales should be consistent and complementary with each other as the authors make use of the same atmosphere models to obtain their results. We derived absolute K magnitudes from theoretical luminosities by using the bolometric correction as a function of spectral type (i.e. color, T_{eff}) given in Leggett et al. (2000) down to mid-M classes, and in Reid et al. (2001b) for cooler types. To summarize, the second order polynomial fits ($1700 \leq T_{\text{eff}} \leq 3500$ K) we used are as follows:

$$(J - K) = 6.423 - 3.49 \times 10^{-3} T_{\text{eff}} + 5.41 \times 10^{-7} T_{\text{eff}}^2 \qquad rms = 0.04\,\text{mag}$$

$$BC_K = 5.745 - 1.46 \times 10^{-3} T_{\text{eff}} + 1.57 \times 10^{-7} T_{\text{eff}}^2 \qquad rms = 0.06\,\text{mag}$$

Isochrones in Fig. 6.6 are plotted for $T_{\text{eff}} \leq 2900$ K, which roughly corresponds to masses smaller than $0.1\,M_\odot$ at ages around 1 Gyr.

From Fig. 6.6 we can see that the evolutionary models nicely reproduce the trend delineated by field objects, except for the reddest colors ($J - K > 1.5$) where models apparently predict brighter magnitudes. Of the two sets of isochrones, the Lyon 1–5 Gyr models seem to produce a better fit to the observed data in the field. The difference in color between Gl 569Ba and Gl 569Bb is consistent with the spectral types of the objects. Within $1\,\sigma$ the uncertainties of our JK photometry, the location of the pair is well matched by isochrones in the age interval 0.2–1.0 Gyr. This indicates a young age for the multiple system, a result which is compatible with the elevated X-ray emission of the "single" M2.5-type primary (Pallavicini, Tagliaferri & Stella 1990; Huensch et al. 1999), with the system belonging to the young Galactic disk as inferred from its kinematics (Reid, Hawley & Gizis 1995), with the large rotation rate measured for the star Gl 569A (Marcy & Chen 1992), as well as with the late-M spectral type–lithium–age relationships (Magazzù et al. 1993; see Bildsten et al. 1997).

Lithium is detected in M8–M9 Pleiades BDs (Rebolo et al. 1996; Stauffer et al. 1998), whereas older and slightly more massive objects have depleted it very efficiently. Thus, lithium non-detections in very late-M type objects necessarily imply ages older than the

Pleiades. No lithium feature is observed in the composite optical spectrum of Gl 569Bab (Magazzù et al. 1993), thus implying that the binary is older than 0.12 Gyr. The R-band spectroscopic data shown in Magazzù et al. (1993) have poor signal-to-noise ratio, and are dominated by the bright and more massive component as it contributes twice as much flux as does the fainter component. We will discuss later how the age of the system can be constrained to a much smaller range using results presented in this paper.

We do not find from our near-infrared photometry strong evidence for the possible binary nature of Gl 569Ba as claimed by Martín et al. (2000a) on the basis of their H vs $H - K$ color-magnitude diagram as well as by Kenworthy et al . (2001) on the basis of their $J - K$ colors. The relative position of the two components of the pair in Fig. 6.6 reasonably fits the location of field dwarfs even within $1\,\sigma$ the error bars. If Gl 569Ba is a binary itself, the smaller companion has to be at least a factor five less luminous. We have combined our AO K-band images to look for any possible companion. We place a $3\,\sigma$ limit at $K = 16.5\,\mathrm{mag}$ (0.015–0.02 $M_\odot$) on the brightness of a possible companion at distances greater than 0.25 arcsec, and less than 2 arcsec– half the size of the AO detector. We cannot discard, however, the presence of extremely faint and less massive objects around any of the components which our AO observations have not been able to detect/resolve. Follow-up high resolution spectroscopy and/or very detailed analysis of the orbital motion of the pair may reveal the presence of close-in giant planets.

6.5.2 The HR Diagram and Substellarity

Figures 6.7 and 6.8 show the location of Gl 569Ba and Gl 569Bb in the HR diagram (luminosity as a function of effective temperature) and provide a comparison with state-of-the-art evolutionary models by the Lyon group (Chabrier et al. 2000) and by the Arizona group (Burrows et al. 1997). Solar-metallicity abundance isochrones of ages 120, 300, 500 Myr and 1 Gyr, and evolutionary tracks of masses in the interval 0.04–0.09 $M_\odot$ are shown in these figures. We use the most recent determination of the substellar mass limit at 0.072 $M_\odot$ to define the stellar-substellar borderline. Because this value is the smallest of those available in the literature (see e.g. Grossman, Hays & Graboske 1974; D'Antona & Mazzitelli 1994; Burrows et al. 1993), our conclusions on substellarity will be conservative. We indicate in the figures the substellar mass boundary and the location of the 50% depletion limit of lithium burning predicted by the two sets of models. We have obtained the luminosity and effective temperatures of our targets as explained above. The third-order polynomial fit that gives temperatures as a function of the observed $(J - K)$ color is the following:

$$T_{\text{eff}} = 7744.6 - 9488.4(J - K) + 5509.9(J - K)^2 - 1135.7(J - K)^3 \qquad rms = 50\,\text{K}$$

This fit has been calculated for colors in the range 0.85–2.06 (spectral types M6–L6), and is based on the temperature calibrations provided by Basri et al. (2000) and Leggett et al. (2000). Bolometric corrections in the J- and K-bands from Reid et al. (2001b) and Bessell, Castelli & Plez (1998), respectively, have been used to transform magnitudes into bolometric luminosities. The values we derive for the pair are listed in Table 6.1. The error bars in luminosity take into account the uncertainty of the distance modulus (Hipparcos) and the photometric uncertainties, leading to a total uncertainty of ± 0.07 dex. The error bars assigned to the effective temperatures come from the uncertainty in the colors alone. These are in general a factor 2 larger than the rms of the polynomial fit describing the temperature calibration.

Models should be able to provide explanations to all physical properties so far known for Gl 569Ba and Gl 569Bb, i.e., photometry, the total mass of the pair and the destruction of lithium. From Fig. 6.7 we observe that the location of Gl 569Ba is consistent with severe lithium depletion, in agreement with available optical spectroscopic observations. According to the Lyon models, even the fainter companion Gl 569Bb has depleted its lithium. The likely age of the system is in the range 0.2–1 Gyr, but only with younger ages is it possible to reproduce the astrometric total mass derived for the pair. Therefore, the real constraint to the age of the system is given by the total mass rather than by the error bars in the figure. For the age of 300 Myr, the binary would be formed by objects of 0.069 $M_\odot$ (Gl 569Ba) and 0.059 $M_\odot$ (Gl 569Bb), in good agreement with the mean orbital solution. Both masses are below the stellar-substellar borderline and thus the two components would be brown dwarfs. For slightly older ages, individual masses would become larger, as would the total mass. On the basis of the largest possible astrometric total mass value at the $3\,\sigma$ level, the pair would be made up of an object on the substellar borderline with 0.078 $M_\odot$ and a 0.070 $M_\odot$-BD at the age of 500 Myr. We tabulate the possible masses of the pair as a function of age in Table 6.5; these estimations rely on the Lyon models.

The comparison with the Arizona models shown in Fig. 6.8 also yields very young ages

---

[a]Should have (partially) preserved lithium.

[b]These estimates match the mean astrometric orbital solution.

[c]The total mass inferred for this age is even beyond the $3\,\sigma$ uncertainty of our astrometric solution.

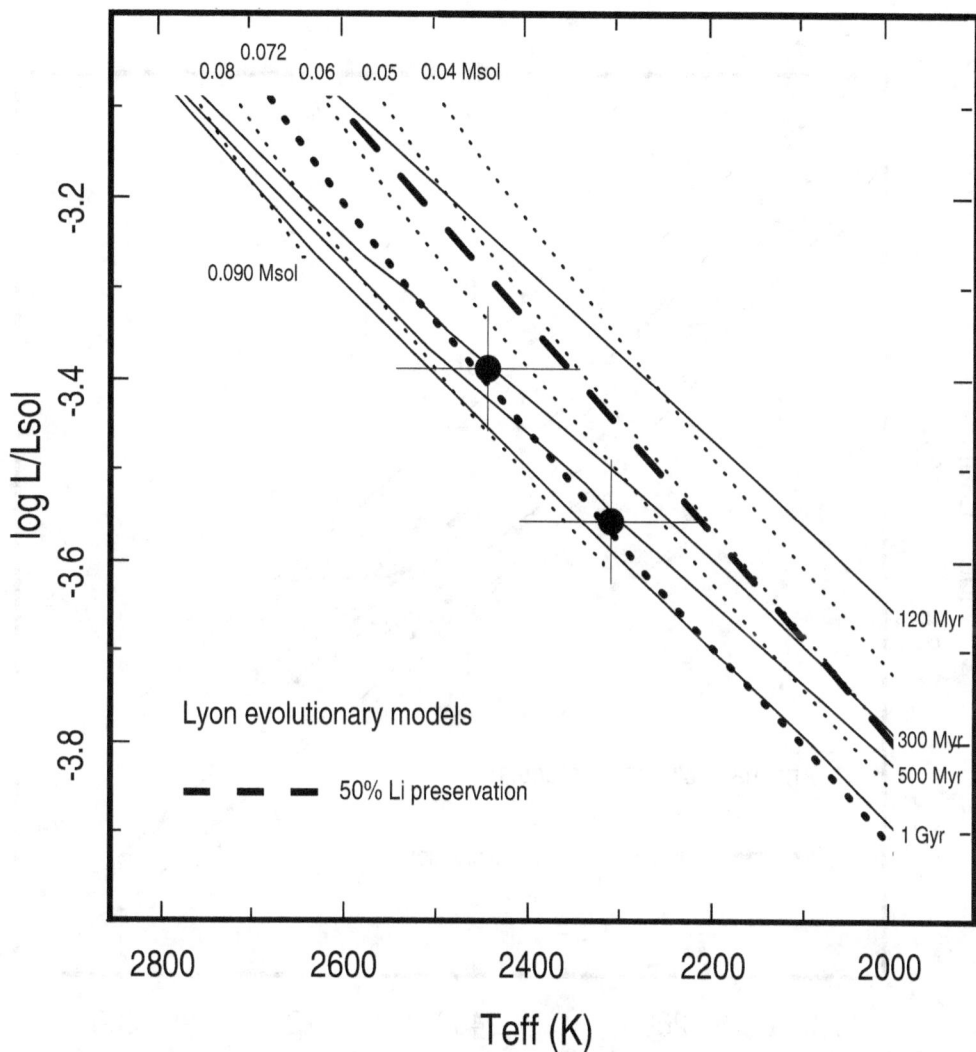

Figure 6.7: HR diagram illustrating the location of Gl 569Ba and Gl 569Bb (filled dots) in comparison with theoretical evolutionary tracks of constant mass (dotted lines) and isochrones (full lines) from the Lyon group (Chabrier et al. 2000). The track corresponding to the substellar mass limit at 0.072 $M_\odot$ is shown with a thicker dotted line. Masses in solar units are labelled on the upper part of the diagram, and ages for the isochrones are indicated to the right. The thick dashed line indicates the 50% lithium depletion limit predicted by the Lyon models. Objects to the left have severely depleted lithium, whereas objects to the right still preserve a significant amount of this element. Solar abundance has been assumed in generating this figure.

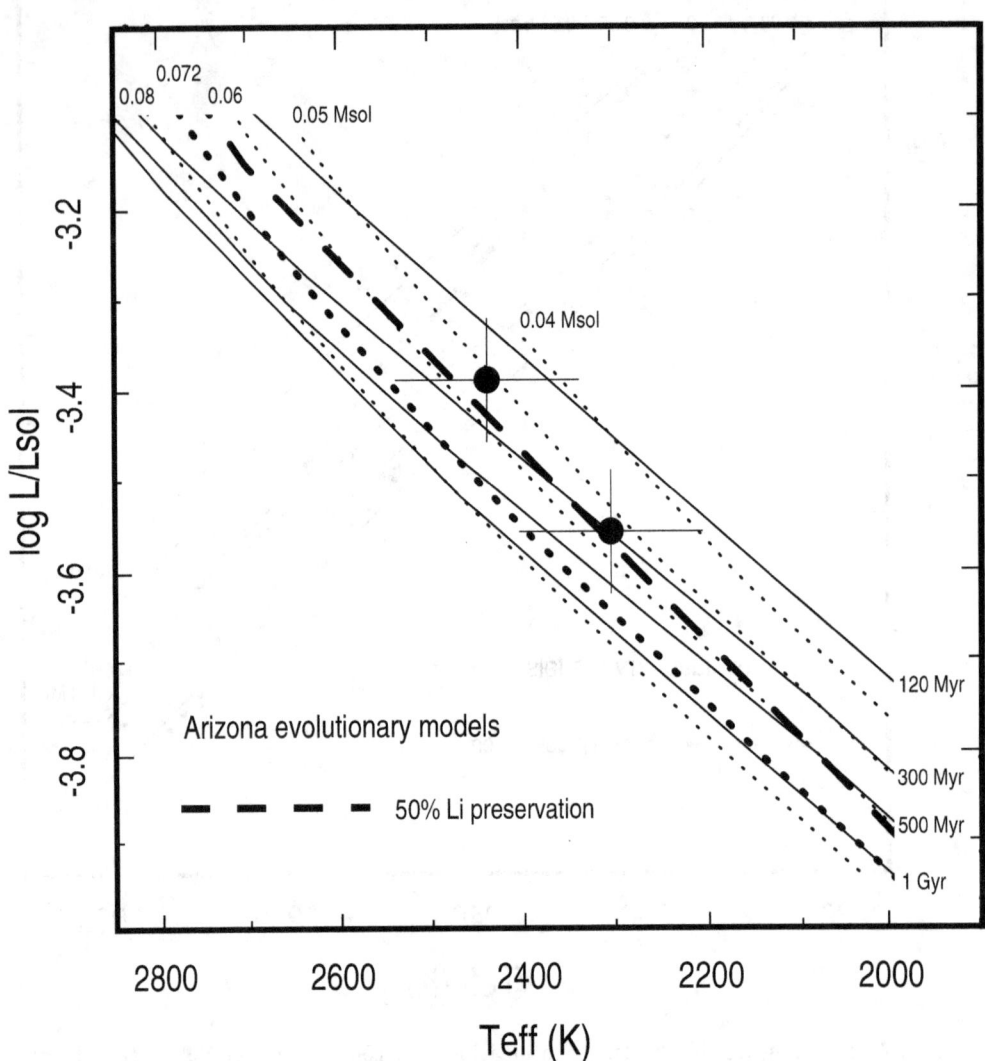

Figure 6.8: HR diagram illustrating the location of Gl 569Ba and Gl 569Bb (filled dots) in comparison with theoretical evolutionary tracks of constant mass (dotted lines) and isochrones (full lines) from the Arizona group (Burrows et al. 1997). See the caption of Fig. 6.7 for further details. Here, the 50% lithium preservation line (thick dashed line) is taken from the Arizona models for consistency.

Age (Myr)	Gl 569Ba ($M_\odot$)	Gl 569Bb ($M_\odot$)	Total mass ($M_\odot$)
200	0.055	0.048[a]	0.103
250	0.061	0.053[a]	0.114
300[b]	0.069[b]	0.059[b]	0.128[b]
400	0.072	0.065	0.137
500	0.078	0.070	0.148
700[b]	0.082[c]	0.074[c]	0.156[c]

Table 6.5: Likely ages and individual masses of Gl 569Ba and Gl 569Bb. Based on the Lyon models (Chabrier et al. 2000; Baraffe et al. 1998). The uncertainty of individual mass models is ±0.002 $M_\odot$.

(120–500 Myr) for the system, and thus also small masses for each of the components. To this point, theory and some observations seem to be consistent. However, the Arizona evolutionary models predict hotter temperatures (by about 100 K) around the substellar mass limit at a given age than do the Lyon models, and hence according to these models both Gl 569Ba and Gl 569Bb should have preserved a considerable amount of lithium in their atmospheres. Only by assuming the highest temperatures allowed by the error bars can the Arizona models account for the observed lithium depletion in Gl 569Bab. This would move the pair to a location between 300 Myr and 500 Myr in Fig. 6.8. In order to be consistent with the additional restriction of the astrometric total mass, ages in the interval 300–500 Myr are required; the resulting masses are 0.055–0.075 $M_\odot$ for Gl 569Ba and of 0.048–0.068 $M_\odot$ for Gl 569Bb.

Collecting evidence from the orbital solution, photometry, spectroscopy, and the comparison with evolutionary models, the most likely scenario of the binary Gl 569Bab is: two very late M-type BDs with masses of 0.055–0.078 $M_\odot$ and 0.048–0.070 $M_\odot$ in a close orbit, in turn orbiting an early M-type 0.5 $M_\odot$ star, the whole system with an age in the range 250–500 Myr. Such young ages found in nearby objects are not surprising since it now seems that the Sun is located close to a region that was the site of substantial amounts of recent stellar formation (Zuckerman & Webb 2000).

Figure 6.9 portrays the mass-luminosity relationship for different ages as given by the evolutionary models. The two members of the pair are plotted with error bars indicating the uncertainty in luminosity and the likely mass range of each component. Masses that have depleted lithium by a factor 2 are also incorporated into the figure. According to the

Arizona models, Gl569Bb may have preserved lithium in its atmosphere, whereas this is quite unlikely based on the Lyon models. Therefore, lithium observations of this BD are needed in order to discriminate which model reproduces neatly the properties of the pair. In addition, precise radial velocity measurements of each component will lead to an accurate determination of the individual masses, and thus will also constitute a better constraint on the models. Nevertheless, models do not appear to be far from reproducing the observational properties of Gl569Bab. Delfosse et al. (2000) show that the mass-luminosity relationship given by the Lyon models reasonably describes the low-mass stellar regime in the field. The pair Gl569Bab has lower masses that belong to the substellar regime. Brown dwarfs around stars have been reported in the recent years (see Table 5 in Reid et al. 2001a for a compilation of the complete list), but to our knowledge none of them has been proved to be a binary itself. Gl569Bab turns out to be the first confirmed resolved binary BD as a companion to a star.

The distance to Gl569 implies a physical separation of 49 AU between the M2.5-type star and the substellar pair. This large separation and the high mass ratio ($q \sim 0.135$ and 0.120) between the star and each of the BDs favors the fragmentation of a self-gravitational collapsing molecular cloud as the most plausible explanation for the formation of the system (Boss 2000; Bodenheimer 1998). Whether each component of the pair Gl569Bab originated from a second fragmentation and collapse process of a small cloud core is not clear (the physical separation is 0.92 AU, and the mass ratio of the pair is $q \sim 0.89$). The activity of the nascent low-mass star when it was gaining mass and becoming more luminous may have caused the disruption of the less massive collapsing core into two close substellar objects before the hydrostatic core could build up enough mass to eventually start hydrogen burning. Energetic outflows and jets up to thousands of AU in length have been detected in low-mass stars of very young star forming regions (e.g., Reipurth et al. 2000; Fridlund & Liseau 1998). We cannot discard the possibility, however, that the protoplanetary disk around the star might have also played an important role in the origin of the companions. Disks extending up to several hundred AU are known to exist around stars (Bruhweiler et al. 1997). Clearly, finding other similar systems will, in addition to providing additional dynamical masses, also contribute to our knowledge of the genesis of such interesting multiple low-mass binaries.

6.6 Conclusions

We have obtained new observations of the Gl569Bab pair (low-mass binary companion at a wide separation from the M2.5-type star Gl569A), which have allowed us to derive the spectral types of each component and the orbital parameters of the system. We find that the total mass of the low-mass binary is $0.125^{+0.023}_{-0.02} M_\odot$ (3-σ) with two detected components of M8.5 and M9 spectral types (half a subclass uncertainty) completing one eccentric ($e = 0.32 \pm 0.01$) orbit every 876 ± 9 days. We have also acquired new J and K near-IR photometry in order to locate Gl569Ba and Gl569Bb in the HR diagram and compare them with the most recent evolutionary models by the Lyon group (Chabrier et al. 2000) and the Arizona group (Burrows et al. 1997). The pair is likely formed by two solar metallicity young brown dwarfs with masses in the interval 0.055–0.078 $M_\odot$ (Gl569Ba) and 0.048–0.070 $M_\odot$ (Gl569Bb) at the young ages of 250–500 Myr. Our adaptive optics images taken with the Keck II telescope exclude the presence of any other resolved companion with K magnitudes brighter than 16.5 (3 σ) at separations of 0.25 arcseconds up to 2 arcseconds from Gl569Bab. This detection limit corresponds to masses around 0.015–0.02 $M_\odot$ for the possible age range of the system. Further radial velocity and astrometric measurements will be very valuable to detect giant planets, as well as to provide individual masses for each of the members of the pair.

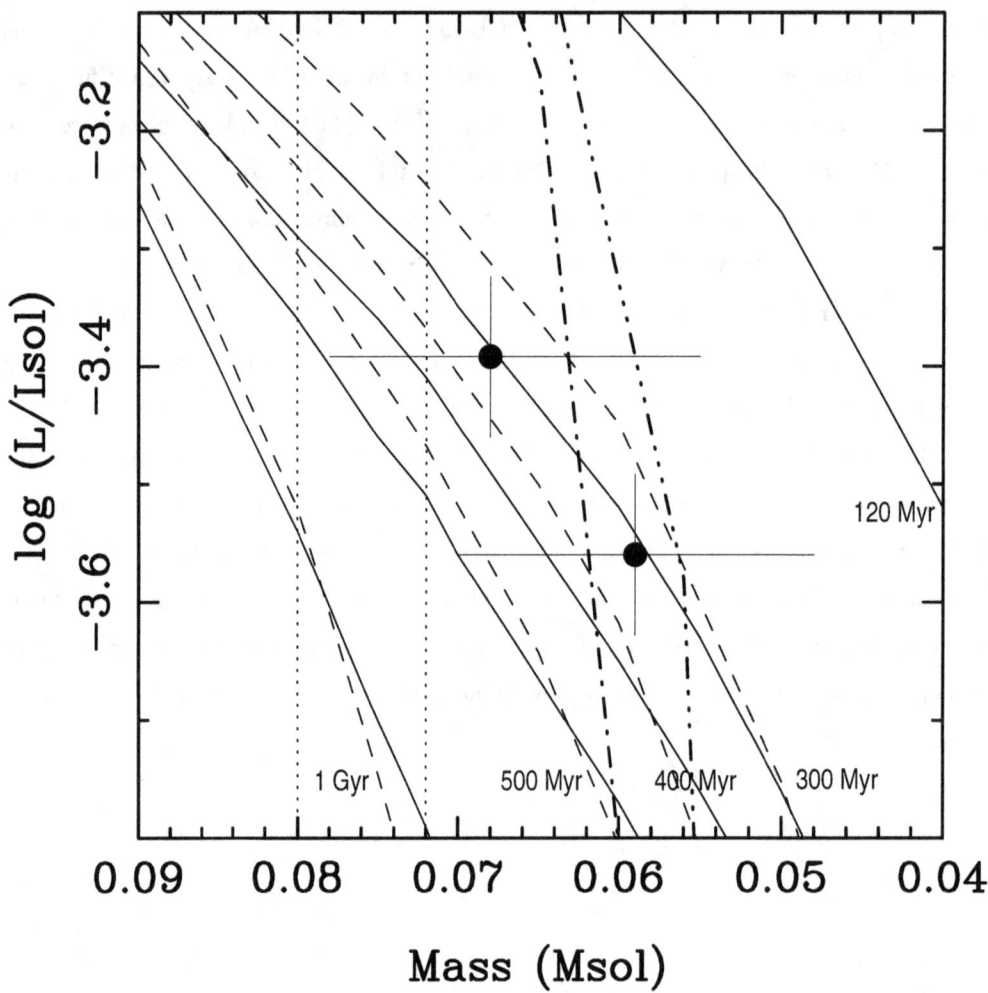

Figure 6.9: Mass-luminosity relationship for different ages (120, 300, 400 and 500 Myr, and 1 Gyr) according to the Lyon models (Chabrier et al. 2000 — full lines) and to the Arizona models (Burrows et al. 1997 — dashed lines). The stellar-substellar transition mass range at 0.072–0.080 $M_\odot$ is indicated by the vertical dotted lines. The dot-dot-dot-dashed line (Lyon) and the dot-dashed line (Arizona) mark when 50% of the lithium has burned. Lithium preservation occurs to the right of these lines. The error bars assigned to Gl 569Ba and Gl 569Bb (filled circles) correspond to the uncertainty in luminosity, and to the likely mass range (3-σ) derived for each object.

Chapter 7

Summary and Future Outlook

7.1   Low-Mass Stars

I have used three experiments to obtain direct measurements of of masses and radii of several low-mass stars. First, I have used PTI to measure the apparent angular diameters of stars in the 0.15–0.8 $M_\odot$ mass range, with a precision of 2–10%. Second, I have used adaptive optics and interferometric observations to resolve two low-mass binary systems, and hence derive dynamical mass estimates. One of the systems, GJ 569B, is the lowest mass system with a dynamical mass determination: with a combined mass of $0.125^{+0.023}_{-0.02}$ solar masses, the system contains at least one, and likely two objects with masses below the substellar limit.

These measurements can be compared to a range of published models, e.g. Baraffe et al. (1998), in a variety of ways such as mass vs. color, or radius vs color. In general, the Baraffe models compare well with the observations, with only small deviations.

The prospect for future observations is bright, as a number of more capable interferometers are in the commissioning stages. Recently, 4 diameter measurements were obtained with the VLTI (Ségransan et al. 2003), and more are promised. In general, increasing the available sample of diameter measurements will require longer baseline systems (~ 200m); as current systems are limited to observing star with diameters larger than ~ 1 mas, which severely limits the number of available targets. Fortunately, both the VLTI and CHARA[a] arrays will have baselines in excess of 200 meters, and are expected to have improved limiting magnitudes compared to PTI. Thus I expect that in the next few years the number of high precision diameter estimates (i.e.,with precisions $\sim 5\%$ or better) should reach a few

---

[a]Center for High Angular Resolution Astronomy, located on Mt. Wilson, CA.

tens.

Once the sample size has reached a dozen or so, it becomes desirable to directly observe the effects of limb darkening, as such corrections can amount to a few percent and hence limb darkening will likely become the dominant source of systematic uncertainty. However, at can be seen in Fig. **??**, disentangling the effects of limb darkening from that of a size change requires observations past the first null of the visibility curve – a challenging prospect. However, as such measurements will be crucial for models of both Cepheids and low mass stars, they should be undertaken. I will further discuss this in Sec. 7.3

In addition to the diameter measurements, I have obtained a small number of dynamical mass estimated for low-mass objects. As such they represent a second area where the models can be tested, and again, it is the models of Baraffe et al. that provide the best comparison. However, in the case of the GJ 569B system, there is some controversy surrounding the models. In particular, in Chapter 6 we found that the system is best fit by a model age of 300 Myr and total mass of 0.128 solar masses. As one might expect for substellar objects (which lack substantial internal energy sources and hence cool off over a few Gyr), increasing the age of the system while keeping the total luminosity fixed at the observed value results in an increased total system mass, something that is inconsistent with the dynamical mass measurement. However, Reid et al. (2002) have argued that the system cannot be as young as 300 Myr, based primarily on an observed X-ray luminosity of GJ 569A that lies between that of Pleiades and Hyades members of the same spectral type. If there is indeed an age discrepancy, the models may have to be revised. Clearly, high precision mass and luminosity measurements of these objects represents a good test of models, and in particular the details of how the luminosity changes with age.

7.2 Cepheids

I have used high precision angular diameter measurements done at PTI to resolve the pulsations of two Galactic Cepheids. Applying a Baade-Wesselink analysis I derive distance estimates accurate to $\sim 10\%$, currently the best available direct distance estimates for individual Cepheids. Although a sample of two is too small to draw any definitive conclusions, the measured Cepheid distances are consistent with previous distances to $\sim 1\sigma$, lending a measure of confidence to the indirect results.

The value of direct Cepheid distances is that they are useful for calibrating the Cepheid period-luminosity relationships that underpin the ladder of techniques used in determining cosmological distance scales; currently there is a ~ 0.1 magnitude systematic uncertainty in such scales. Given that the Baade-Wesselink technique is purely geometrical, it avoids many systematic uncertainties associated with the photometry. Conversely, the measured Cepheid diameters can be used to find surface brightness vs. color relations that can subsequently be used in indirect distance determinations (the Barnes-Evans method, 1976). In fact such surface brightness relations, albeit ones based on non-pulsating stars, have been used for many years as the principal indirect means of determining Cepheid distances. Here too, the agreement between our observations and previous work is good.

The next step in studying Cepheids with interferometry should be, as with the case of low-mass stars, to expand the sample size. Here the next generation of longer-baseline systems will be useful, although unlike the case with low-mass stars, there are a number of (comparatively) large but faint Cepheids – the faintness being due to interstellar extinction. Hence it may be the case that improved limiting magnitudes would be more profitable than longer baselines; in such a case the KI and VLTI interferometers should prove particularly useful.

However, regardless of how small the statistical uncertainties can be made there remains a substantial level of systematic uncertainty ($\sim 5 - 10\%$). This uncertainty is due to our limited understanding of Cepheid atmospheres, in particular the details of limb darkening and its close relative, the radial velocity projection factor. However, a small number of observations of Cepheid limb darkening would serve to constrain existing atmospheric models and drastically reduce this systematic error. This is where interferometry can provide uniquely valuable information.

7.3   Limb-Darkening Observations

Observing limb-darkening with an interferometer is difficult for two reasons. First, as stated earlier the degeneracy between size and limb darkening is only broken for $\pi B\theta/\lambda > \sim 3.8$, thus long baselines are required; although not inherently particularly difficult, such long baselines have until now not been available. More of a concern is the fact that observations

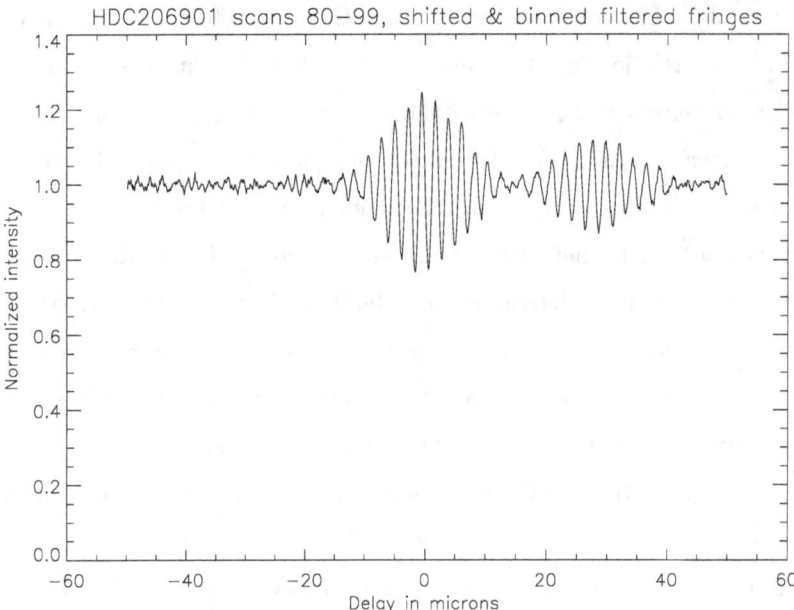

Figure 7.1: A example of self-phase referencing. Using the primary fringe tracker to stabilize the OPD, we scan through the zero relative delay position and measure the intensity as a function of delay. In the case of a close (0.1 arcsecond) binary star the result is a double fringe pattern; this can be used to obtain very high precision relative astrometry between the componenents.

near or past the first visibility null have of necessity a low SNR [b] as SNR is proportional to fringe amplitude; recall Fig. 1.8. However, given the ability to combine three interferometric baselines at a time one can choose two of the baselines to be relatively short and hence have a high SNR. The third baseline can be up to twice as long as the others, providing twice the resolution. Although the SNR on the long baseline is lower, the fact that the baselines form a closed loop implies that by tracking the fringe location on the two high-SNR baselines one automatically knows the location of the fringes on the low-SNR baseline, and can thus integrate coherently for long periods of time. This technique is perfect for measuring stellar limb darkening; however, it requires a suitable interferometer such as CHARA or NPOI.

7.4 Phase Referencing

I have demonstrated phase referencing with a fringe-tracking interferometer, which allows one to use an isoplanatic reference star to sense and correct atmospherically induced OPD fluctuations in real time. Such corrections allows the use of longer exposure times; potentially up to several seconds, as compared to 10–20 ms without phase referencing. I have modeled the phase referencing servo system in order to predict the amount of residual OPD variation and the resulting fringe smearing. I find that for a system like PTI, with a suitable choice of servo parameters, the loss in fringe visibility can be limited to $\sim 30\%$, while a higher bandwidth system results in correspondingly improved performance.

Phase referencing is of critical importance to the planned Keck Outrigger astrometry program, where current designs call for the use of astrometric references as faint as 5 magnitudes fainter than the primary star. As a result of the successful demonstration of both phase referencing and narrow angle astrometry at PTI, the Keck Interferometer has been built and should begin an astrometry program in the near future.

In addition to its use in astrometry, phase referencing can be used to phase[c] an interferometric array so that the stabilized beams can be fed to other instruments; because the beams are stabilized, these other instruments are not limited to the short exposure times required for fringe tracking. This allows a great deal of flexibility in the instrument design, e.g., the use of high-resolution spectrographs, or multi-way beam combination. One example where this might be desirable is the case of an imaging interferometer: as stated before, a good image quality requires extensive coverage of the uv plane, and hence many apertures and baselines. However, in a direct-detection interferometer combining many apertures requires either splitting the light from each aperture many ways, or all-in-one combination with separate phase modulation applied to each beam. In the first case, the presence of read noise in the detector sets a limit to the number of splits before the SNR drops below 1, while in the second case photon noise from larger apertures can overwhelm the signal from smaller apertures (or longer baselines). The result is that many-way beam combination is inherently inefficient, and if such a beam combiner was limited to short exposure times the limiting magnitude of the array would be pathetic. Therefore it is desirable to split the two functions into separate systems using phase referencing.

---

[b]In the immortal words of D. Mozurkewich: "Your fringes can be good, or they can be interesting."

[c]i.e., control the OPD fluctuations.

Another use of phase referencing is illustrated in Fig.7.1. Here, phase referencing is used to stabilize the beam, while a second delay line+beam combiner scans through the relative path delay. The resulting fringe pattern can be measured in order to obtain very high precision relative astrometry of the components; or, in the case of single stars, can be Fourier-transformed and hence provide a measure of fringe visibility as a function of wavelength. This latter mode is referred to as double-Fourier interferometry, and represents an interesting new way of simultaneously obtaining high spatial and spectral resolution information.

Appendix A

Fringe Tracking and Servos

A.1 Control Theory Review

For a complete introduction or review of control theory, the reader is referred to e.g.
Franklin, Powell & Emami-Naeini (1994); here we briefly review the details relevant to
a fringe tracking servo system. Consider a simple servo system, consisting of an actuator
and a sensor, that operates on a single output parameter, and is affected by some external
disturbance. How does one control the actuator such that the result is as close as possible
to the desired input? The obvious answer is to arrange the sensor such that it measures
the output of the actuator, compares that output with the desired input, and adjusts the
command sent to the actuator accordingly. Such a system is a "feedback" system, and is
illustrated in Fig. A.1.

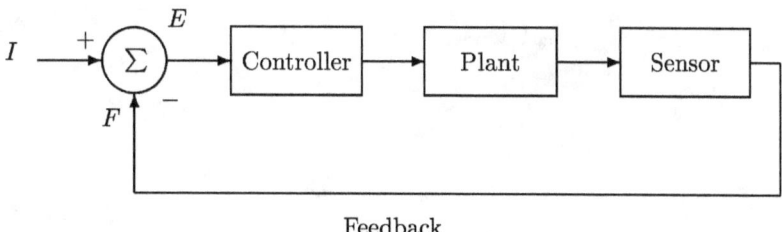

Figure A.1: A simple control loop.

How does the error, E, i.e. the difference between the command input, I, and the system response, F, behave? Let the total system response to an error signal E be given by some gain function, G (this includes the effect of the controller, plant and sensor combined, for simplicity). Hence

$$F = GE \qquad (A.1)$$

The error function itself is simply

$$E = I - F \qquad (A.2)$$

giving

$$\tilde{E} \equiv \frac{E}{I} = \frac{1}{1+G} \qquad (A.3)$$

Generally the system response function is caluated as a function of frequency, using one-sided Laplace transforms, i.e.

$$F(s) = \int_0^\infty f(t)e^{-st}dt \qquad (A.4)$$

where s is a complex variable. In calculating the frequency response of a servo system, I take

$$s = j2\pi f \qquad (A.5)$$

Below I derive the frequency response for a few common servo system components.

A.1.0.1 Time Delay

If a component of the servo system introduces a time delay of T we have

$$x_{out}(t) = x_{in}(t - T) \qquad (A.6)$$

which gives

$$X_{out}(s) = \int_0^\infty x_{in}(t - T)e^{-sT}dt \qquad (A.7)$$

$$= e^{-sT}X_{in}(s) \qquad (A.8)$$

If we define the gain function as $G(s) \equiv X_{out}(s)/X_{in}(s)$ we find

$$G_{delay}(f, T) = e^{-j2\pi fT} \tag{A.9}$$

A.1.0.2 Integrator

A common form of servo controller uses an integrator, thus

$$x_{out}(t) = 2\pi f_c \int_0^t x_{in}(t')dt' \tag{A.10}$$

where f_c corresponds to the closed-loop bandwidth of the servo for the case where the plant gain is near unity. Assuming $x_{in}(t \leq 0) = 0$ and $\lim_{t \to \infty} e^{-st} \int_0^t x_{in}(t')dt' = 0$ integration by parts yields

$$
\begin{aligned}
X_{out}(s) &= \int_0^\infty x_{out}(t)e^{-st}dt &\text{(A.11)}\\
&= 2\pi f_c \int_0^\infty \int_0^t x_{in}(t')e^{-st}dt'dt &\text{(A.12)}\\
&= 2\pi f_c \frac{1}{s} \int_0^\infty x_{in}(t')e^{-st'}dt' &\text{(A.13)}\\
&= 2\pi f_c \frac{X_{in}(s)}{s} &\text{(A.14)}
\end{aligned}
$$

or

$$G_f(f) = -j\frac{f_c}{f} \tag{A.15}$$

A.1.0.3 Sampling

A digital control system uses discrete sampling, while in this treatment I am approximating the system response as continuous. The effect of discrete sampling can be approximated as

$$x_{out}(t) = \int_{t-T_s/2}^{t+T_s/2} x_{in}(t')dt' \tag{A.16}$$

where T_s is the sampling time. The same approach as in Sec. A.1.0.2 yields

$$X_{out}(s) = \frac{1}{s}X_{in}(s)\left[e^{sT_s/2} - e^{-sT_s/2}\right] \tag{A.17}$$

and hence

$$G_{Sampling}(f) = \frac{\sin(\pi f T_s)}{\pi f T_s} \tag{A.18}$$

A.1.1 On Servo Stability

From Eqn. A.3 it is clear that one must avoid any situation where $G(j2\pi f) = -1$, else the servo will be unstable; in practice this limits the servo gain.

A.2 Integrating Servo

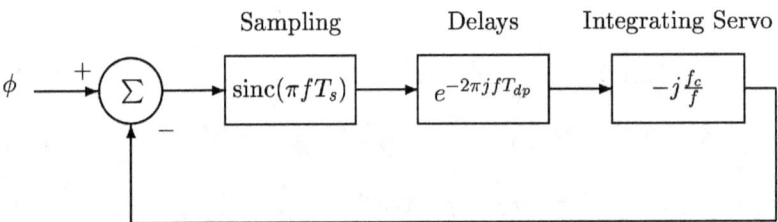

Figure A.2: The PTI primary fringe tracker operates a simple feedback loop based on an integrating servo.

The error rejection of the feedback loop for this sampled-data system can be approximated as

$$\tilde{E}_1(f) \quad = \quad \frac{1}{1+G} \tag{A.19}$$

$$= \quad \frac{1}{1 - j\frac{f_c}{f}\mathrm{sinc}(\pi f T_s)\mathrm{e}^{-j2\pi f T_{dp}}} \tag{A.20}$$

The PSD filter function $H_{fb}(f)$ follows from

$$H_{fb}(f) \quad = \quad \tilde{E}_1(f)\tilde{E}_1^*(f) \tag{A.21}$$

$$= \quad \left(\frac{1}{1 - j\frac{f_c}{f}\mathrm{sinc}(\pi f T_s)\mathrm{e}^{-j2\pi f T_{dp}}} \right) \left(\frac{1}{1 + j\frac{f_c}{f}\mathrm{sinc}(\pi f T_s)\mathrm{e}^{+j2\pi f T_{dp}}} \right) \tag{A.22}$$

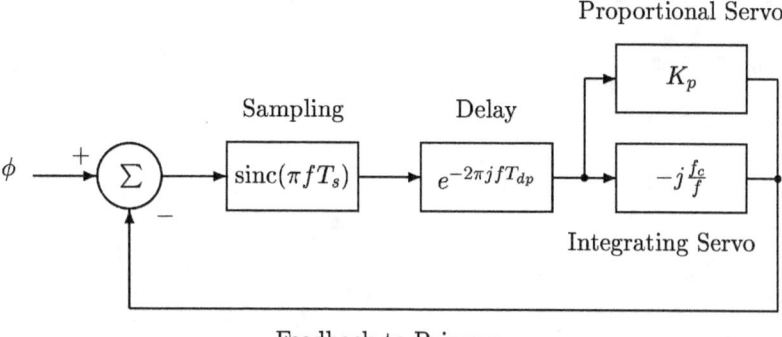

Figure A.3: The PTI primary fringe tracker updated to include a proportional gain term for improved performance.

$$= \frac{1}{1 - 2\frac{f_c}{f}\mathrm{sinc}(\pi f T_s)\sin(2\pi f T_{dp}) + \left(\frac{f_c}{f}\right)^2 \mathrm{sinc}^2(\pi f T_s)} \tag{A.23}$$

A.3   Integral-Proportional Servo

Although the simple integrating servo works quite well, it is desirable to further improve the error rejection for improved phase referencing performance. Adding a proportional gain to the control loop (Fig.A.3) accomplishes this by reducing the peaking near f_c.

$$\tilde{E}_{pi}(f) = \frac{1}{1+G} \tag{A.24}$$

$$= \frac{1}{1 + (K_p - j\frac{f_c}{f})\mathrm{sinc}(\pi f T_s)\mathrm{e}^{-j2\pi f T_{dp}}} \tag{A.25}$$

hence the transfer function is

$$H_{pi}(f) = \tilde{E}_{pi}(f)\tilde{E}_{pi}^*(f) \tag{A.26}$$

$$= \left(\frac{1}{1 + (K_p - j\frac{f_c}{f})\mathrm{sinc}(\pi f T_s)\mathrm{e}^{-j2\pi f T_{dp}}}\right)\left(\frac{1}{1 + (K_p + j\frac{f_c}{f})\mathrm{sinc}(\pi f T_s)\mathrm{e}^{+j2\pi f T_{dp}}}\right) \tag{A.27}$$

$$= \frac{1}{1 + 2\mathrm{sinc}(\pi f T_s)(K_p \cos(2\pi f T_{dp}) + \frac{f_c}{f}\sin(2\pi f T_{dp})) + (K_p^2 + \left(\frac{f_c}{f}\right)^2)\mathrm{sinc}^2(\pi f T_s)} \tag{A.28}$$

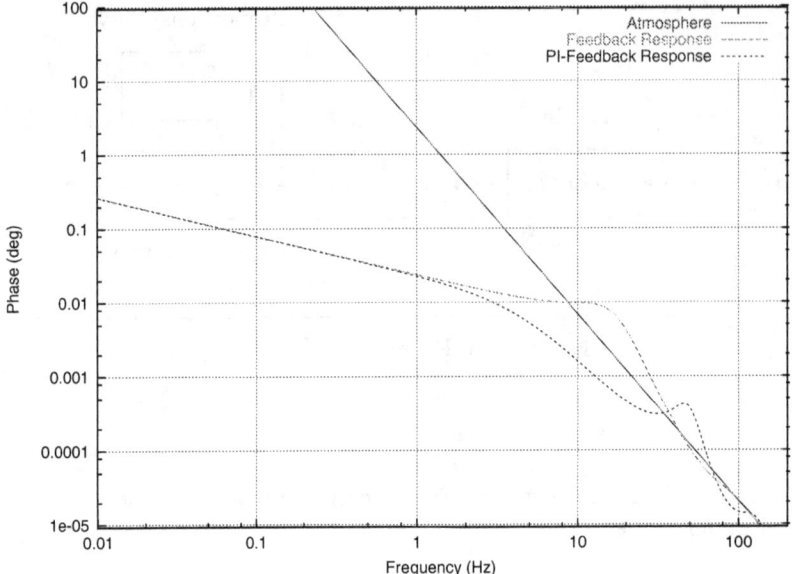

Figure A.4: This is what H_{fb} and H_{pi} look like, when multipled by an atmospheric power law $\phi \propto f^{-2.58}$. Model parameters were $f_c = 10$ Hz, $T_{dp} = 11$ ms, $T_s = 6.75$ ms, $K_p = 0.5$.

The resulting fringe tracker error PSD, for both integral and proportional-integral control, is shown in Fig.A.4.

A.4 Integrating Servo with Feedforward

As discussed in Chapter 5, in the case of phase referencing the usual gain limitation required for servo stability does not apply, as the secondary delay line is not part of the feedback loop (see Fig.A.5). Hence one can apply the full measured error to the secondary side. By inspection

$$O_1 = I - E_1 \tag{A.29}$$

$$O_2 = O_1 + E_1\text{sinc}(\pi f T_s)e^{-j2\pi f T_{ds}} \tag{A.30}$$

$$= I - E_1 + E_1\text{sinc}(\pi f T_s)e^{-j2\pi f T_{ds}} \tag{A.31}$$

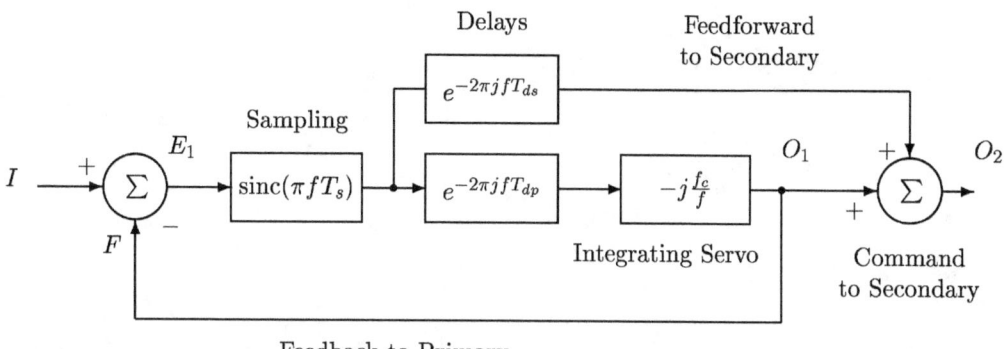

Figure A.5: The PTI control loop used in phase referencing. E_1 is the primary side servo error, O_1 is the primary side servo output (i.e. the motion of the primary delay line). O_2 is the secondary side output.

define the secondary error function as

$$E_2 \equiv I - O_2 \tag{A.32}$$

$$= E_1(1 - \mathrm{sinc}(\pi f T_s)e^{-j2\pi f T_{ds}}) \tag{A.33}$$

or

$$\tilde{E}_2(f) = \tilde{E}_1(f)\left(1 - \mathrm{sinc}(\pi f T_s)e^{-j2\pi f T_{ds}}\right) \tag{A.34}$$

Thus $\tilde{E}_2(f)$ is the product of the feedback servo rejection function and a simple time-delay limited rejection function. When the feedback servo gain goes to 0 ($f_c \to 0$), $\tilde{E}_1(f) \to$ 1 and the response is just the time-delay limited response. The PSD filter function is

$$H_{ff}(f) = \tilde{E}_2(f)\tilde{E}_2^*(f) \tag{A.35}$$

$$= \tilde{E}_1(f)\tilde{E}_1^*(f)\left(1 - \mathrm{sinc}(\pi f T_s)e^{-j2\pi f T_{ds}}\right)\left(1 - \mathrm{sinc}(\pi f T_s)e^{+j2\pi f T_{ds}}\right) \tag{A.36}$$

$$= H_{fb}(f)\left(1 - 2\mathrm{sinc}(\pi f T_s)\cos(2\pi f T_{ds}) + \mathrm{sinc}^2(\pi f T_s)\right) \tag{A.37}$$

The resulting servo error PSD is shown in Fig.A.6. The steep decline in servo error at low frequencies allows long integration times while maintaining a high fringe visibility.

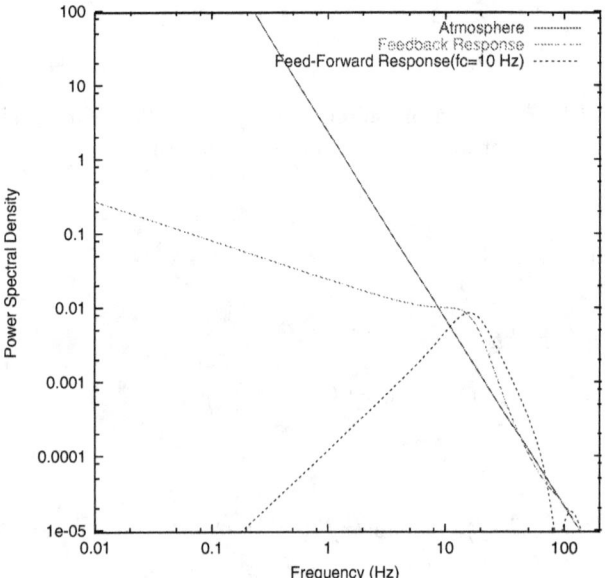

Figure A.6: The effect of feedforward. Plotted is H_{fb} and H_{ff} multipled by an atmospheric power law $\phi \propto f^{-2.58}$. Model parameters were $f_c = 10$ Hz, $T_{dp} = T_{ds} = 11$ ms, $T_s = 6.75$ ms.

Bibliography

[Albrow & Cottrell 1994] Albrow, M. D., Cottrell, P. L., 1994, MNRAS, 267, 548–556

[Allen 1982] Allen, C. W., 1982, Astrophysical Quantities, Athlone Press, London, U.K.

[Allen 2000] Allen, C. W. 2000, Allen's Astrophysical Quantitities. Fourth edition, ed. Arthur N. Cox, New York: Springer Verlag, p. 151

[Allard et al.1997] Allard, F., Hauschildt, P. H., Alexander, D. R., Starrfield, S., 1997, Ann. Rev. Astr. Ap., 35, 137.

[Andersen 1991] Andersen, J., 1991, Astron. Astropys. Rev., 3, 91.

[Armstrong et al. 2001] Armstrong, J. T., Nordgren, T. E., Germain, M. E., Hajian, A. R., Hindsley, R. B., Hummel, C. A., Mozurkewich, D., Thessin, R. N. 2001, AJ, 121, 476

[Armstrong et al. 2001b] Armstrong, J. T. et al. 2001, American Astronomical Society Meeting, 198, 6303

[Baraffe & Chabrier 1996] Baraffe, I., Chabrier, 1996, ApJ, 461, L51.

[Baraffe et al. 1998] Baraffe, I., Chabrier, G., Allard, F., & Hauchildt, P. H. 1998, A&A, 337, 403.

[Barnes & Evans 1976] Barnes, T. G. & Evans, D. S. 1976, MNRAS, 174, 489.

[Barnes et al. 1997] Barnes, T. G., Fernley, J. A., Frueh, M. L., Navas, J. G., Moffett, T. J., Skillen, I., 1997, PASP, 109,645–658.

[Basri et al. 1996] Basri, G., Marcy, Geoffrey. W., & Graham, J. R. 1996, ApJ, 458, 600

[Basri et al. 2000] Basri, G., Mohanty, S., Allard, F., Hauschildt, P. H., Delfosse, X., Martín, E. L., Forveille, T., & Goldman, B. 2000, ApJ, 538, 363

[Becklin & Zuckerman 1988]   Becklin, E. E., & Zuckerman, B. 1988, Nature, 336, 656

[Bersier et al. 1994]   Bersier, D., Burki, G., Mayor, M., Duquennoy, A., 1994, A&A, 108, 25–39

[Bersier 2002]   Bersier, D., submitted to ApJ.

[Bersier et al. 1997]   Bersier, D., Burki, G., Kurucz, R. L., 1997, A&A, 320, 228–236

[Bessell et al. 1998]   Bessell, M. S., Castelli, F., & Plez, B. 1998, A&A, 333, 231

[Bildsten et al. 1997]   Bildsten, L., Brown, E. F., Matzner, C. D., & Ushomirsky, G. 1997, ApJ, 482, 442

[Boden et al. 1998]   Boden, A. F., van Belle, G. T., Colavita, M. M., Dumont, P. J., Gubler, J., Koresko, C. D., Kulkarni, S. R., Lane, B. F., Mobley, D. W., Shao, M., Wallace, J. K., 1998, ApJl, 505, L39

[Boden et al. 1999a]   Boden, A. et al. 1999a, ApJ, 515, 356.

[Boden et al. 1999b]   Boden, A. et al. 1999b, ApJ, 527, 360.

[Boden et al. 2000]   Boden, A., Creech-Eakman, M., and Queloz, D., 2000, ApJ, 536, 880-890.

[Bodenheimer 1998]   Bodenheimer, P. 1998, Asp. Conf. Ser., 134, p. 115

[Bohm-Vitense & Proffitt 1985]   Bohm-Vitense, E. , Proffitt, C. 1985, ApJ, 296, 175–184.

[Bono, Caputo, & Marconi 1998]   Bono, G., Caputo, F., and Marconi, M. 1998, ApJL, 497, L43

[Bopp & Evans 1973]   Bopp, B., Evans, D. 1973 (BE73), MNRAS, 164, 343.

[Bopp & Fekel 1977]   Bopp, B., Fekel, F. 1973, AJ, 82, 490.

[Bopp et al. 1980]   Bopp, B., Noah, P., and Klimke, AJ, 85, 1386.

[Boss 2000]   Boss, A. P. 2000, ApJ, 545, L61

[Bruhweiler et al. 1997]   Bruhweiler, F., et al. 1997, Bull. Am. Astron. Soc., 191, 47.03

[ten Brummelaar et al. 2001] ten Brummelaar, T. A. et al. 2001, American Astronomical Society Meeting, 198, 6106

[Burrows et al. 1993] Burrows, A., Hubbard, W. B., Saumon, D., & Lunine J. I. 1993, ApJ, 406, 158

[Burrows 1997] Burrows, A., et al. 1997, ApJ, 491, 856

[Butler et al. 1999] Butler, R. P., Marcy, G. W., Fischer, D. A., Brown, T. M., Contos, A. R., Korzennik, S. G., Nisensen, P., Noyes, R. W., 1999, ApJ, 526, 916–927.

[Casali & Hawarden 1992] Casali, M. M., Hawarden, T. 1992, UKIRT Newsletter, 4, 33

[Chabrier & Baraffe 1995] Chabrier, G., Baraffe, I., 1995, ApJl, 451, L29.

[Chabrier et al. 2000] Chabrier, G., Baraffe, I., Allard, F., Hauschildt, P. 2000, ApJ, 542, 464

[Chabrier & Baraffe 2000] Chabrier, G., Baraffe, I. 2000, Ann. Rev. Astron. Astrophys., 38, 337.

[Charbonneau et al. 2000] Charbonneau, D. et al., 2000, ApJL, 529, L45–L48.

[Claret et al. 1995] Claret, A., Diaz-Cordoves, J., Gimenez, A., 1994, A&AS, 114, 247-252.

[Clemens et al. 1998] Clemens, J. C., Reid, I. N.,Gizis, J. E., O'Brien, M.S., 1998, , 496, 352.

[Cochran et al. 1997] Cochran W., et al., 1997,ApJ, 483, 457

[Cohen et al. 1999] Cohen, M., Walker, R., Carter, B., Hammersley, P, Kidger, M., Noguchi, K., 1999, AJ, 117, 1864-1889.

[Colavita et al. 1991] Colavita M. M., Hines, B. E., Shao, M., Klose, G. J., Gibson, B. V., 1991, Proc. SPIE, 1542, 205

[Colavita et al. 1992] Colavita, M. M., Hines, B. E., Shao, M., 1992, in ESO Conf. and Workshop Proc. 39, High Resolution Imaging by Interferometry II, ed. F. Merkle (Garching: ESO), 1143

[Colavita et al. 1994] Colavita, M. M., et al., 1994, Proc. SPIE, 2200, 89

[Colavita et al. 1999]   Colavita, M. M., Wallace, J. K., Hines, B. E., Gursel, Y., Malbet, F., Palmer, D. L., Pan, X. P., Shao, M. , Yu, J. W., Boden, A. F., Dumont, P. J., Gubler, J. Koresko, C. D., Kulkarni, S. R., Lane, B. F., Mobley, D. W., van Belle, G. T, 1999, ApJ, 510 , 505–521.

[Colavita 1999]   Colavita, M. M., 1999, PASP, 111, 111–117.

[D'Antona & Mazzitelli 1994]   D'Antona, F., Mazzitelli, I. 1994, ApJs, 90, 467

[Delfosse et al. 1999] Delfosse, X., Forveille, T., Mayor, M., Burnet, M., Perrier, C., 1999, A&A, 341, L63.

[Delfosse et al. 2000]   Delfosse, X., Forveille, T., Ségransan, D., Beuzit, J.-L., Udry, S., Perrier, C., Mayor, M. 2000, A&A, 364, 217

[Eisner et al. 2002]   Eisner, J., Lane, B. F., Sargent A., Hillenbrand, L., 2002, ApJ in press.

[ESA1997]   ESA, 1997, The Hipparcos catalogue, SP-1200

[Evans & Jiang 1993]   Evans, N. R., Jiang, J. H., 1993, AJ, 106, 726–733.

[Faber 1995]   Faber, T. E., Fluid Dynamics for Physicists, 1st Ed., Cambridge Univ. Press, 1995.

[Feast 1999]   Feast, M. 1999, PASP, 111, 775

[Feast & Catchpole 1997]   Feast, M. W. & Catchpole, R. M. 1997, MNRAS, 286, L1

[Fernie 1984]   Fernie, J. D. 1984, ApJ, 282, 641

[Fernie 1990]   Fernie, J. D., 1990, ApJs, 72, 153–162.

[Fernley, Skillen, & Jameson 1989]   Fernley, J. A., Skillen, I., Jameson, R. F. 1989, MN-RAS, 237, 947

[Festin 1998]   Festin, L. 1998, A&A, 333, 497

[Fizeau 1868]   Fizeau, H., 1868, C. R. Acad. Sci, 66, 932–934.

[Forrest, Skrustskie & Shure 1988]   Forrest, W. J., Skrustskie, M. F., Shure, M. 1988, ApJ, 330, L119

[Fouque & Gieren 1997] Fouque, P., Gieren, W.P., 1997, A&A, 320, 799-810.

[Franklin, Powell, & Emami-Naeini 1994] Franklin, G., Powell, D., Emami-Naeini, A., "Feedback Control of Dynamical Ststems", 3rd Ed, Addison-Wesley, 1994.

[Fridlund & Liseau 1998] Fridlund, C. V. M., Liseau, R. 1998, ApJ, 499, L75

[Gieren 1988] Gieren, W. P. 1988, ApJ, 329, 790.

[Gieren, Moffett, & Barnes 1999] Gieren, W. P., Moffett, T. J., Barnes, T. G. 1999, ApJ, 512, 553

[Gizis, Kirkpatrick & Wilson 2001] Gizis, J. E., Kirkpatrick, J. D., Wilson, J. C. 2001, AJ, in press

[Glebocki & Stawikowski 1995] Glebocki, R., Stawikowski, A. 1995, AcA 45, 725.

[Glebocki & Stawikowski 1997] Glebocki, R., Stawikowski, A. 1997, A&A, 328, 579.

[Grossman, Hays & Graboske 1974] Grossman, A. S., Hays D., & Graboske H. C. 1974, A&A, 30, 95

[Hajian et al. 1998] Hajian, A. et al., 1998, ApJ, 496, 484.

[Hall 1986] Hall, D. 1986, ApJ, 309, L83.

[Handbury-Brown & Twiss 1956] Hanbury-Brown, R., Twiss, R., 1956, Nature, 177, 27.

[Hanbury-Brown et al. 1974] Hanbury-Brown, R., Davis, J., Allen, L. R., 1974, MNRAS, 167, 121.

[Heintz 1991] Heintz, W. D. 1991, AJ, 101, 1071

[Henry & Kirkpatrick 1990] Henry, T. J., Kirkpatrick, J. D. 1990, ApJ, 354, L29

[Henry et al. 2000] Henry, T. J., Benedict, G. F., Gies, D. R., Golimowski, D. A., Ianna, P. A., Mason, B. D., McArthur, B. E., Nelan, E. P., Torres, G., 2000, "MASSIF: Masses and Stellar Systems with Interferometry", SIM Key Project, http://sim.jpl.nasa.gov

[Henry & McCarthy 1993] Henry, T.J., McCarthy, D. W., 1993, AJ, 106, 773.

[Hindsley & Bell 1986] Hindsley, R., Bell, R. A., 1986, PASP, 98, 881–888

[Hodgkin et al. 1999]   Hodgkin, S. T., Pinfield, D. J., Jameson, R. F., Steele, I. A.,Cossburn, M. R., Hambly, N. C. 1999, MNRAS, 310, 87

[Huensch et al. 1999]   Huensch, M., Schmitt, J. H. M. M., Sterzik, M. F., Voges, W. 1999, A&A Supp., 135, 319

[Hut 1981]   Hut, P. 1981, A&A, 99, 126.

[Jacobsen & Wallerstein 1981]   Jacobsen, T. S., Wallerstein, G., 1981, PASP, 93, 481–485.

[Jacobsen & Wallerstein 1987]   Jacobsen, T. S., Wallerstein, G., 1987, PASP, 99, 138–140.

[Jacoby et al. 1992]   Jacoby, G. H. et al. 1992, PASP, 104, 599

[Jones et al. 1994]   Jones, H. R. A., Longmore, A. J., Jameson, R. F., Mountain, C. M. 1994, MNRAS, 267, 413

[Jones et al. 1996]   Jones, H. R. A., Longmore, A. J., Allard, F., Hauschildt, P. H. 1996, MNRAS, 280, 77

[Jones & Tsuji 1997]   Jones, H.R.A., Tsuji, T. , 1997, /apjl, 480 L39.

[Kenworthy et al. 2001]   Kenworthy, M., et al., 2001, ApJL, 554, L67–L70.

[Kervella et al. 2001]   Kervella, P., Coudé du Foresto, V., Perrin, G., Schöller, M., Traub, W. A., Lacasse, M. G. 2001, A&A, 367, 876

[Kirkpatrick et al. 1991]   Kirkpatrick, J. D., Henry, T. J., McCarthy, D. W. Jr. 1991, ApJs, 77, 417

[Kirkpatrick, & McCarthy 1994]   Kirkpatrick, J. D., McCarthy, D. W. Jr. 1994, AJ, 107, 333

[Kirkpatrick et al. 1999]   Kirkpatrick, J. D., et al. 1999, ApJ, 519, 802

[Kirkpatrick et al. 2000]   Kirkpatrick, J. D., et al. 2000, AJ, 120, 447

[Kirkpatrick et al. 2001]   Kirkpatrick, J. D., Dahn, C. C., Monet, D. G., Reid, I. N., Gizis, J. E., Liebert, J., Burgasser, A.J. 2001, AJ, 121, 323.

[Krockenberger, Sasselov, & Noyes 1997] Krockenberger, M., Sasselov, D. D., Noyes, R. W. 1997, ApJ, 479, 875

[Labeyrie 1975] Labeyrie, A., 1975, AJ, 196, L71.

[Lane et al. 2000] Lane, B. F., Kuchner, M. J., Boden, A. F., Creech-Eakman, M., Kulkarni, R. R., 2000, Nature, 407, 485–487.

[Laney & Stobie 1995] Laney, C. D., Stobie, R. S., 1995, MNRAS, 274, 337–360

[Laughlin & Adams 1999] Laughlin, G., Adams, F. C., 1999, ApJ, 526, 881-889.

[Lawson 1997] Lawson, P. R. 1997, Selected papers in Long Baseline Stellar Interferometry, SPIE Milestone Series, Vol. MS 139, SPIE Press, Bellingham WA.

[Lawson 2001] Lawson, P. R. 2001, Principles of Long Baseline Interferometry, http://sim.jpl.nasa.gov/library/coursenotes.html

[Leggett, Allard & Hauschildt 1998] Leggett, S. K., Allard, F., Hauschildt, P. H. 1998, ApJ, 509, 836

[Leggett et al. 2000] Leggett, S. K., Allard, F., Dahn, C., Hauschildt, P. H., Kerr, T. H., Rayner, J. 2000, ApJ, 535, 965

[Leinert et al. 2000] Leinert, C., Jahreiß,H., Woitas, J., Zucker, S., Mazeh, T., Eckart, A., Köhler, R. 2001, A&A, 367, 183

[Leung & Schneider 1978] Leung, K., Schneider, D. P., 1978, AJ, 83, 618.

[Linfield et al. 1999] Linfield, R. P., et al., 1999, Working on the Fringe: An International Conference on Optical and IR Interferometry from Ground and Space, Dana Point, CA, May 24-27, 1999. ASP Conference Series (S. Unwin and R. Stachnik, editors), p. 58.

[Lloyd, Oppenheime & Graham 2002] Lloyd, J. P., Oppenheimer, B.R., Graham, J.R., 2002, Pub. Astron. Soc. Aus. , 19, Issue 3, 318–322.

[Lucke & Mayor 1980] Lucke, P. and Mayor, M. 1980 (LM80), A&A, 92, 182.

[Magazzù, Martín & Rebolo 1993] Magazzù, A., Martín, E. L., & Rebolo, R. 1993, ApJ, 404, L17

[Mandel & Wolf 1995] Mandel, L., Wolf. E., 1995, Optical Coherence and Quantum Optics, Cambridge Univ. Press.

[Marcy & Chen 1992] Marcy, G. W., Chen, G. H. 1992, ApJ, 390, 550

[Martín et al. 1998] Martín, E. L., Basri, G., Gallegos, J. E., Rebolo, R., Zapatero Osorio, M. R., Béjar, V. J. S. 1998, ApJ, 499, L61

[Martín, Basri & Zapatero Osorio 1999] Martín, E. L., Basri, G., Zapatero Osorio, M. R. 1999, AJ, 118, 1005.

[Martín et al. 2000a] Martín, E. L., Koresko, C. D., Kulkarni, S. R., Lane, B. F., Wizinowich, P. L.,2000a, ApJ, 529, L37.

[Martín et al. 2000b] Martín, E. L., Brandner, W. Bouvier, J., Luhman, K. L., Stauffer, J. R., Basri, G., Zapatero Osorio, M. R., & Barrado y Navascués, D. 2000b, ApJ, 543, 299

[Mayor & Queloz 1995] Mayor, M., Queloz, D., 1995, Nature, 6555, 355.

[McLean et al. 1998] McLean, I.S., et al. 1998, Proc. SPIE, 3354, 566

[McLean et al. 2000] McLean, I. S., et al. 2000, ApJ, 533, L45

[Metcalfe et al. 1996] Metcalfe, T. S., Mathieu, R. D., Latham, D. W., Torres, G., 1996, ApJ, 456, 356.

[Michelson 1891] Michelson, A., 1981, Nature, 45, 160–161.

[Michelson & Pease 1921] Michelson, A., Pease, F., 1921, AJ, 53, 249–259.

[Moffett & Barnes 1984] Moffett, T. J. & Barnes, T. G. 1984, ApJs, 55, 389–432.

[Moffett & Barnes 1987] Moffett, T. J. & Barnes, T. J. 1987, ApJ, 323, 280

[Nordgren et al. 2000] Nordgren, T., Armstrong, J. T., Germain, M. E., Hindsley, R. B., Haijan, A. R., Sudol, J. J., Hummel, C. A., 2000, ApJ, 543, 972–978.

[Nordgren et al. 2001] Nordgren, T., Lane, B. F., Hindsley R. B., Kervella, P., 2002, AJ, 123, 3380.

[Pallavicini, Tagliaferri & Stella 1990] Pallavicini, R., Tagliaferri, G., Stella, L. 1990, A&A, 228, 403

[Percy 1993] Percy, J., 1993, PASP, 105, 1422–1426.

[Perley et al. 1989] Perley, R. et al., 1989, "Synthesis Imaging in Radio Astronomy", NRAO Workshop No. 21, ASP Conference Series, Vol. 6.

[Perryman et al. 1997] Perryman, M. et al., 1997, A&A, 323, L49

[Press et al. 1986] Press, W.H., Flannery, B.P., Teukolsky, S.A., Vetterling, W.T., 1986, Numerical Recipies (New York: Cambridge Univ. Press).

[Quirrenbach et al. 1994] Quirrenbach, A.,et al., 1994, A&A, 286, 1019

[Rebolo et al. 1996] Rebolo, R., Martín, E. L., Basri, G., Marcy, G., Zapatero Osorio, M. R. 1996, ApJ, 469, L53

[Reid, Hawley & Gizis 1995] Reid, I. N., Hawley, S. E., Gizis, J. E. 1995, AJ, 110, 1838

[Reid & Gizis 1997] Reid, I.N., Gizis, J.E., 1997, AJ, 113, 2246.

[Reid et al. 2000] Reid, I. N., Kirkpatrick, J. D., Gizis, J. E., Dahn, C. C., Monet, D. G., Williams, R. J., Liebert, J., Burgasser, A. J. 2000, AJ, 119, 369

[Reid et al. 2001a] Reid, I. N., Gizis, J. E., Kirkpatrick, J. D., Koerner, D. W. 2001a, AJ, 121, 489

[Reid et al. 2001b] Reid, I. N., Burgasser, A. J., Cruz, K. L., Kirkpatrick, J. D., Gizis, J. E. 2001b, AJ, in press

[Reid et al. 2002] Reid, N. et al, 2002, AJ, 124, 519–540.

[Reipurth et al. 2000] Reipurth, B., Yu, K. C., Heathcote, S., Bally, J., Rodríguez, L. F. 2000, AJ, 120, 1449

[Ripepi et al. 1997] Ripepi, V., Barone, F., Milano, L., Russo, G., 1997, A&A, 318, 797–804

[Sabbey et al. 1995] Sabbey, C.N., Sasselov, D.D., Fieldus, M.S., Lester, J.B., Venn, K.A., Butler, R.P., 1995, ApJ, 446, 250–260

[Sasselov & Lester 1990]   Sasselov, D. D., Lester, J. B., 1990, ApJ, 362, 333–345.

[Schwab & Cotton 1983]   Schwab, F. R., Cotton, W. D. 1983, AJ, 88, 688–694.

[Schwarzschild 1896]   Schwarzschild, K., 1896, Asrton. Nachr. 139, No. 3335.

[Segransen et al. 2003]   Ségransan, D, Kervella, P., Forveille, T., Queloz, D., Astron. Astrophys 397, L5-L8 2003.

[Shao & Staelin 1980]   Shao, M., Staelin, D., 1980, Appl. Opt., 19, 1519.

[Shao & Colavita 1992a]   Shao, M., Colavita, M. M., 1992a, Ann. Rev. Astron.& Astrophys., 30, 457

[Shao & Colavita 1992b]   Shao, M., Colavita, M. M., 1992b, A&A, 262, 353

[Stauffer, Schultz & Kirkpatrick 1998]   Stauffer, J. R., Schultz, G., Kirkpatrick, J. D. 1998, 499, L199

[Stéphan 1873]   Stéphan, E., 1873, C. R. Acad. Sci., 76, 1008–1010.

[Stéphan 1874]   Stéphan, E., 1874, C. R. Acad. Sci., 78, 1008–1012.

[Szabados 1991]   Szabados, L., 1991, Comm. Konkoly Obs., No. 96.

[Tanvir1999]   Tanvir, N. R. 1999, ASSL Vol. 237: Post-Hipparcos cosmic candles , 17

[Tinney et al. 1997]   Tinney, C. G., Delfosse, X., Forveille, T. 1997, ApJ, 490, L95

[Vogt & Fekel 1979]   Vogt, S. and Fekel, F. 1979 (VF79), ApJ, 234, 958.

[Welch 1994]   Welch, D. L., 1994, AJ, 108, 1421–1426

[Wesselink 1946]   Wesselink, A. J., 1946, BAN, 368, 91

[Wisniewski & Johnson 1968]   Wisniewski, W. Z., Johnson, H. L. 1968, Communications of the Lunar and Planetary Laboratory, 7, 57

[Wizinowich et al. 1998]   Wizinowich, P. L., et al. 1998, Proc. SPIE, 3353, 568.

[Zboril & Byrne 1998]   Zboril, M., Byrne, P. B. 1998, MNRAS, 299, 753

[Zuckerman & Webb 2000]   Zuckerman, B., Webb, R. A. 2000, ApJ, 535, 959

[Zuckerman et al. 1997] Zuckerman, B. et al. 1997, AJ, 114, 805.

www.ingramcontent.com/pod-product-compliance
Lightning Source LLC
Chambersburg PA
CBHW081123170526
45165CB00008B/2528